LONELY MINDS IN THE UNIVERSE

T0220567

LONELY
MINDS
IN THE
UNIVERSE

Giancarlo Genta

Copernicus Books
An Imprint of Springer Science+Business Media

In Association with
Praxis Publishing Ltd

Giancarlo Genta
Department of Mechanics
Politecnico di Torino
Torino 101291
Italy
www.giancarlogenta.it

© 2010 Praxis Publishing, Ltd.

Published in the United States by Copernicus Books,
an imprint of Springer Science+Business Media.

Copernicus Books
Springer Science+Business Media
233 Spring Street
New York, NY 10013
www.springer.com

ISBN 978-1-4419-2225-0   e-ISBN 978-0-387-69039-1

Manufactured in the United States of America.
Printed on acid-free paper.

9 8 7 6 5 4 3 2 1

To Franca and Alessandro

CONTENTS

Contents

Contents

PREFACE

GALILEO discovered in 1610 with his new telescope the four big satellites orbiting Jupiter. Johannes Kepler, informed of the great event, wrote a long and enthusiastic letter to Galileo, the so-called *Dissertatio cum Nuncio Sidereo.* In this opus he is referring to the discovery of the satellites as proof that planet Jupiter is inhabited:

> Si enim quatuor Planetae Iovem circumcursitant disparibus intervallis et temporibus: quaeritur cui bono, si nulli sunt in Iovis globo, qui admirandam hanc varietate suis notent oculis? Nam quod nos in hac terra attinet, nescio quibus rationibus quis mihi persuadeat, ut illos nobis potissimum servire credam, qui illos nunquam conspicimus; neque est expectandum, ut tuis Galilaee ocularibus universi instructi, illos porro vulgo observaturi simus.[1]

And his conclusion:

> ... quatuor hos novos non primario nobis in Tellure versantibus, sed proculdubio Iovialibus creaturis, globum Iovis circumhabitantibus comparatos.[2]

This is probably one of the first mentions of extraterrestrials in the present sense of the word. Planets were not considered habitable worlds during previous centuries, but only points of light in the sky, and alien beings were imagined mostly as folkloristic monsters.

---

[1] Therefore, if four planets orbit Jupiter at different distances and times: one asks to the benefit of whom, if nobody is on planet Jupiter to admire this variety with his eyes? Then, for what we are concerned with on this Earth, I wonder; for what convincing reason? Above all, how can they be useful to us who never see them; and we do not expect that everybody can use their eye-pieces to observe them.

[2] The new four [planets] are not primarily for us who live on the Earth but without doubt for the creatures who live on Jupiter.

But Kepler's arguments were accepted unaltered until the twentieth century, when astronomy and space research gradually discovered that the planetary bodies of the Solar System—except Earth itself—are not the property of intelligent beings. Even independent microbial life has not yet been discovered on the Moon or on the planets. Astronomy during the last century enormously widened the limits of the observed Universe, surrounding an absolutely insignificant, tiny Earth. Billions of stars, nebulae, galaxies, and clusters of galaxies populate the empty space into billions of light-years outward. But there is as yet no proof that life as we know it exists anywhere else in this enormous, diversified Universe.

On the other hand, life on Earth exists everywhere from the depths of the crust up to the stratosphere. Recent discoveries have demonstrated that the hidden microbial part of life might represent the predominant majority of the biomass. Our Earth, our Solar System, and our Galaxy are simple components of the Universe, and there is no evidence that they are exceptional objects in any respect.

This is certainly a very serious assertion, probably one of the most important scientific problems of the twenty-first century. *Lonely Minds in the Universe* is a fascinating analysis of this controversial situation with all its important social, philosophical, and even theological implications. In spite of the technical background of its author, the book is an enjoyable read for every thoughful person who is interested in the past, present, and future destiny of humanity. It raises important questions, but many answers are simply not yet available. The author is convinced that the scientific search for extraterrestrial life and intelligence in the Universe deserves every effort because the result will deeply influence our future. I recommend you read this book with an open mind, which will enrich you with new ideas.

Budapest, 2005   Prof. Iván Almár
                        Member and former co-chairman
                        of the IAA Search for
                        Extraterrestrial Intelligence (SETI) Committee

ACKNOWLEDGMENTS

I am deeply indebted to all my colleagues and the experts in the search for extraterrestrial intelligence (SETI), astrobiology, and space exploration, whose suggestions, criticisms, and bibliographical lists were essential in writing this book. In particular, I would like to thank (in alphabetical order) Ivan Almár, Gregory Benford, Frank Drake, Gregory Matloff, Stelio Montebugnoli, Salvatore Santoli, Paul Shuch, Giuseppe Tanzella Nitti, and Giovanni Vulpetti. Their help was invaluable, but the book reflects strictly my ideas and my views on the subject. The responsibility for any errors or omissions is fully mine.

Piero Galeotti and Danilo Noventa, who founded the Italian SETI Study Center with me, shared the certainty that SETI is mostly an interdisciplinary subject and encouraged me to look to the historical, philosophical, and diplomatic aspects of the discipline.

Tuvia Fogel, literary agent but above all friend, spent countless hours revising the English text, and to him I am deeply grateful.

Last, but far from least, this book could not have been written without the support, encouragement, criticisms, and suggestions of my wife Franca—adviser, critic, editor, companion, and best friend for 35 years—who as usual read the manuscript several times.

A note on the illlustrations
I have made every effort to seek permission from the original copyright holders of the figures, and I apologize if there are any cases where I have not been able to achieve my objective. This applies in particular to Figures 3.7, 3.16, 4.9, 4.10, and 4.11.

INTRODUCTION

HUMANKIND has only gradually become aware of its surrounding Universe, and the widening of its perspectives—from the immediate neighborhood of its dwellings to the immensities of the cosmos as we now conceive it—was a slow process. But as soon as it realized the vastness of the Universe, humankind began to wonder whether Earth is an island of life in an ocean of inanimate matter or whether other living beings, perhaps intelligent, conscious ones, dwell in the vastness of the Universe. This question has not yet found a final answer.

It is nevertheless a question of a different kind from the other "basic" questions—those on the purpose and meaning of life—since in this case science may be able to supply an answer. In the past, discussions of how common life is in the Universe, and whether extraterrestrial intelligence exists, were mostly of a philosophical or theological nature, but at present this matter is the subject of scientific investigations that may lead in the future to scientifically certain answers.

The discoveries of planetology in the 1970s and 1980s, mostly due to the advances in space science and technology, led many scientists to think that life on Earth is a unique and nonrepeatable accident, in a Universe that is not only indifferent but generally hostile to life. However, this pessimistic view has recently started to change, and the opinion that life is a common phenomenon, or even a necessary result of the evolution of inanimate matter, is again gaining momentum. And with it comes the hope that humans can, in the future, with their very presence, bring life to the Solar System and beyond.

Life, however, is a very general term, which on Earth includes a large variety of living things, from bacteria to humans, not to mention forms like viruses, which cannot be classified as living organisms or as nonliving matter. It is likely that if extraterrestrial life forms are ever discovered, we will realize that the variety of living things is even greater, perhaps incomparably greater, to the extent that in some cases it will be difficult to determine whether a newly discovered object can be considered to be a living thing or not. There is, however, no doubt that what we are most interested in is finding beings who share with us a particular aspect of life: intelligence.

Intelligence is often considered to be the final stage in the evolution of living matter, and we wonder whether the process that resulted in the appearance of an intelligent and self-conscious species on this planet is an inevitable evolutionary trend or an accident. We also wonder whether intelligence and self-awareness necessarily coincide, or whether it is possible to conceive of a being that possesses one of these characteristics but not the other. Perhaps the presence of both is a mistake in the evolutionary process, which evolution itself—or, as pessimists claim, humanity's tendency to self-destruction—will soon correct.

If it is unlikely that science will, in the near future, provide an answer to the question on extraterrestrial life, the situation regarding intelligent life is even more complex. We now realize that if other intelligent beings exist, they are without doubt at a great distance from us. This rules out the possibility of studying intelligent beings close up, at least for the time being, by sending automatic probes or personally going there. The only possibility is to observe the Universe, searching for signs of their presence.

The activities aimed at discovering extraterrestrial intelligent forms of life and establishing contact with them—generally subsumed under the acronym SETI (*Search for ExtraTerrestrial Intelligence*)—are mostly based on the use of radiotelescopes, though there are scientists who pursue what is known as optical SETI, looking for signs of extraterrestrial intelligence with telescopes.

But while science proceeds with painstaking caution, public opinion is continually bombarded by messages of every kind: extraterrestrials of all types and origins are said to visit our planet, secretly entertaining relationships of various kinds with humans; traces of their passage are said to be present in historical and mythological sources, to the extent that the present (and the past) of our species is said to be shaped by events that occurred far from our planet, even hundreds or thousands of light-years away. And obviously—according to these claims—Earth's scientists and politicians are well aware of all this, but bound to secrecy by a conspiracy that is motivated by public interest (that is, avoiding waves of panic), by the greed for power, or by even more sinister motivations.

These messages may be mystifications, lies that can often be easily unmasked, but their popularity and influence on the public is nevertheless great. Those who believe in a conspiracy (and they are surprisingly numerous) can easily be induced to believe that evidence against something is part of the machination to deny the existence of the conspiracy itself. In other cases the ideas are passed on in good faith; the lower the cultural level and the confidence in science, the easier it is for people to deceive themselves. We must also remember that we are living in times of general disorientation and mistrust in humanity's rational faculties—a time of quick changes in which we are bombarded every day with promises of fantastic novelties and threats of terrible dangers. It may be difficult to distinguish between reality and imagination, and dreams (or, more often, nightmares) may become a substitute for facts.

Unfortunately, even some scientists play a role in these shenanigans. Sometimes they are the first victims of their own imagination, forgetting what Richard Feynman said in 1974: "You must not fool yourselves— and you are the easiest person for you to fool." At other times they more or less earnestly take for granted things that are only such in their dreams, causing disorientation and confusing matters they should be striving to clarify.

In theory science should be safe from this type of thing; when, at the end of the middle ages, the foundations of modern science were laid, William of Occam, in an attempt to ground philosophy in reason and empirical verification, stated the well-known principle now known as *Occam's razor*. It states that when searching for an explanation no non-essential entity must be postulated. In particular, when explaining a phenomenon one must resort to the simplest explanation, the one requiring the postulation of the smallest number of unknown entities. William of Occam also considered it impossible to demonstrate any finality of the Universe.

One criterion that science has always followed is that exceptional discoveries require exceptional evidence. Clearly this cautious practice can hinder new ideas from being accepted and make scientific revolutions more difficult, but science also needs stability and certainties. In particular, observations or deductions that cause a paradigm shift must always be looked at with suspicion. By this we do not mean that scientists must be conservative at all costs, and must fear new ideas; on the contrary, they must always be open-minded and aware that even the most accredited scientific theories are only models and cannot be regarded as absolute, eternal truths. Statements or theories are scientific to the extent that they may be *falsified*; that is, it is possible to contrive an experiment capable of showing that they are false. In true science it is possible to demonstrate that a statement is false, but not that it is true.

Newtonian mechanics, for instance, has for a long time been a good example of a scientific theory—and in fact it still is. From the moment Newton proposed it, he and other scientists thought up and performed a large number of experiments aimed at showing that it was false, but for centuries it withstood all these tests. At the end of the nineteenth century, some experimental results were shown to be in disagreement with the theory's predictions, implying that some refinements of the theory were needed. So relativistic and quantum mechanics, each with its own field of application, were devised. The new theories, which include Newtonian mechanics as a particular case but transcend it, were themselves subjected to experimental verifications that could falsify them, but they proved to be true. There is no doubt that in the future they too will find limits to their scope of application and will in turn be replaced by new theories.

Scientists, therefore, must be open-minded when evaluating results that may invalidate consolidated theories and avoid defending old theories for the sake of conservatism, but they must also try to maintain a balance between cautiousness and mental openness. It is to be expected that true scientific revolutions, those that involve a real paradigm shift, will require many years to be completed and will often have to wait for a new generation of scientists to take over.

But the opposite attitude is also dangerous: to embrace new theories in an indiscriminate way—often after hasty interpretations of dubious experimental results—or to accept as a verified result what is simply a hope, may have serious consequences. It is not just bad science, which may risk the spread of incorrect results or outright errors, but also it may generate quarrels that often degenerate into personal controversies, loaded with emotional aspects that have nothing to do with science. In this situation it happens that an idea or a theory becomes a banner, and the fight to make it prevail becomes more important than the rightful search for the truth. The way is thus open for improper practices, which may go as far as the manipulation of the results of experiments or the fabrication of false evidence.

This can even be done in good faith: the scientist believes in the correctness of a theory with such intensity that when the experiments that should confirm it do not yield the hoped-for results, he or she almost unconsciously adjusts them to fit the direction they were expected to follow. In other cases (perhaps the majority) the scientist may act in bad faith: he has invested so much of his own reputation, or gone so far out on a limb in his statements (usually to obtain funding to gather evidence for his "truth"), that he can no longer back out and is thus forced to proceed at the cost of falsifying evidence.

At the root of these cases is perhaps a lack of critical thought on consolidated theories, but above all on the alternative theories that promise easy explanations. That and the desire to be a protagonist: it is easier to get the media's attention with theories and statements that excite public opinion than with proper, painstaking research work. Gianbattista Marino, at the beginning of the seventeenth century, wrote: "È del poeta il fin la meraviglia; chi non sa far stupir vada alla striglia" (the poet's prize is a wondering gaze; let him scrub floors who can't amaze); nothing is worse than a scientist who indulges in the same attitude.

The criterion of exceptional evidence is therefore an essential antidote to these dangers, and the risk that its strict application may slow scientific progress is a price that has to be paid. If a scientist starts to deviate from these principles, he or she risks drifting into the kingdom of pseudo-science, where statements are made that cannot be demonstrated (or that have been shown many times to be false but are constantly repeated in slightly different forms), statements that may flatly oppose science, but that in some cases claim to contribute to it. Often they are called "alternative science" (obviously alternative to some "official science"), sometimes boasting of a glorious past (at times with good reason, as in the case of astrology or alchemy) and claiming an equally magnificent future. Their hold on public opinion is often strong, as shown by the great number of astrologists and magicians practicing various forms of commercial witch-craft, or by the enviable incomes of many practitioners of "alternative medicine."

The topics discussed in this book are particularly suited to be dealt with by the "alternative sciences," to the point that for serious scientists it has been difficult to speak of extraterrestrial life and, even more so, of extraterrestrial intelligence. As a consequence, a scientist who ventures into these subjects must proceed with great caution, to avoid falling into that no-man's-land where uncertain things are explained by even more uncertain theories, in a spiraling sequence of affirmations that are neither demonstrable nor falsifiable.

Two characteristics emerged as science developed and took shape over the last centuries: reductionism and specialization. Reductionism is basically the subdivision of complex problems into their elementary aspects, each of which can then be approached independently.

This approach has allowed science to face simple problems, often in a strongly idealized form, that could successfully be solved. The specialist who deals with the aspect of a problem in which he or she has specialized, creates a model of a real world in which only those aspects that are relevant for the solution of the idealized problem exist. He or she is not expected to study or be an expert in the disciplines involved in the other aspects of the problem. Even less is he or she expected to deal with them. This formulation is so rooted in the way of thinking of scientists that often they no longer perceive the fact that what they are studying is only a partial (sometimes marginal) aspect of a far more complex reality. Indeed, their ability is often more in isolating the aspect of reality that is relevant for the problem under examination than in reaching a solution. The reductionist formulation spread from theoretical to applied sciences, and was allegedly responsible for most of the striking technological successes of the last few centuries.

Recently, problems were encountered in several fields of science and technology that could not be studied properly in a strictly reductionist way. Strong criticisms of reductionism were also put forward, even outside the field of science, at times to the point of blaming it for most of the evils in our society. Unfortunately, this assault on reductionism often becomes a destructive criticism of science as a whole, or at least of "Western science" (to which an elusive "Eastern science" is opposed, supposedly free from this evil and not as "inhuman") and above all of Western technology. A "holistic science" now opposes "reductionist science." It is based on the assumption that a complex system is fundamentally more than the sum of its parts, and cannot be studied one piece at a time. These positions come, at least in part, from the study of nonlinear systems, for which the principle of the superposition of effects, typical of linear systems, does not hold. In other words, the behavior of a system subject to complex conditions cannot be obtained by subjecting it to the various perturbations one at a time and then adding the results.

The example of celestial dynamics is clarifying: the behavior of each planet can be calculated considering only the planet under examination and the Sun. This is the classical approach taken since Newton's time. But to obtain greater precision, it is essential to account for the perturbations that the planets exert on each other, entering into the field of what is known as "nonlinear astrodynamics," in which the Solar System is studied as a whole. When systems as complex as living beings are studied, in which the whole system is clearly much more than the simple sum of its parts, reductionism often becomes insufficient and many aspects of the problem must be accounted for at the same time.

The problem is that, in this way, when the complexity of a problem grows to the point that its rigorous study becomes impossible, the holistic approach becomes a breech through which pseudo-sciences may creep in and statements that cannot be demonstrated may be elevated from subjective impressions to the rank of scientific results. In the search for extraterrestrial life and intelligence, this danger is always present because the complexity of the subject is great and the disciplines involved are so varied that a single researcher cannot possibly master the whole field. Finally, the emotional impact of the subject is such that true objectivity is often difficult to maintain.

One of the principles that have slowly made their appearance in modern science is the so-called *principle of mediocrity*. The Universe of ancient philosophy was much smaller than the Universe that modern science has revealed to us (though it probably seemed incredibly great to the people of those times), and the Earth and humanity were given the central place in it. It is true that Greek philosophy formulated the heliocentric hypothesis long before Copernicus (Chapter 1), but few philosophers really believed that Earth revolved around the Sun. The Copernican revolution actually consisted in removing humanity from the center of the Universe. It gave life to a completely new way of understanding the Universe and the role of humanity, and also had major consequences outside science. It is not hard to see why such a paradigm shift was so heavily opposed and was accepted only after many years.

But Copernicus had just made the first step. After Earth, the Sun also lost its central place in the Universe. There were precursors here too, from some Greek philosophers to Giordano Bruno, who believed in an infinite Universe without any center, but only modern astronomy could supply evidence for what otherwise was just a hypothesis. The Milky Way we see in the sky was understood to be the edge of the galactic disk, seen from the inside, with the Sun being one of the many stars that orbit in a complex pattern around its center. Then it was time for the galactic center to lose its central position: the Milky Way was recognized to be just one of the billions of galaxies randomly distributed throughout the Universe. Giordano Bruno was almost right: the Universe does not have a center, even if it is very likely not infinite. Modern astrophysics suggests hypothetical views of the Universe that are even further from what common sense (which essentially still advocates a geocentric view, with Earth fixed at the center of a small Ptolemaic Universe) suggests: an infinity of universes (a multiverse), with our Universe being just one of them.

Earth is therefore nothing more than *an average* planet, orbiting around *an average* star, in *an average* galaxy, belonging to *an average* local group, in (perhaps) *an average* universe. But this principle of mediocrity doesn't just apply to space: it holds for time also.

If the Universe has neither origin nor end (as is the case in the *steady-*

state theory, which has little credit today), the present would be just one instant in an endless duration, without any particular meaning. The principle of mediocrity would be complete in time too. Nowadays the most accepted theory on the beginning of the Universe is that of the *Big Bang* (Chapter 3), according to which the Universe had a beginning about 12 or 15 billion years ago, and evolves by expanding toward a very distant future. The present instant can thus be placed in a well-determined phase of cosmic evolution, but this does not really confer on it any peculiar characteristics that might constitute a serious exception to the principle of mediocrity. We, at any rate, live in a phase without any peculiar characteristics of the evolution of a probably still very young Universe. If there is any ground for theories suggesting that a large number of universes are continuously born, in a sort of higher level steady state, the principle of mediocrity would be true in an even more complete way.

We must expressly note that the principle of mediocrity doesn't directly imply that our planet is one of a large number of inhabited planets and that our species is one of many intelligent species that are born, develop, and conclude their existence in this Universe. Consider, for instance, a scenario in which life (and even more so, intelligence) is so rare as to have occurred only once in the past and future history of the Universe. Obviously the only planet that has intelligent life is Earth, and we are the only intelligent species. Even this scenario doesn't violate the principle of mediocrity: if only one intelligent species exists, the only possible conscious observer (us) cannot exist in any other place or time except Earth today. We therefore must be very cautious when invoking the principle of mediocrity to demonstrate the existence of many inhabited planets and many intelligent species.

But if other living or intelligent beings exist, how similar to us will they be? One of the greatest problems facing those who study the possibility of extraterrestrial life is the tendency to anthropomorphism, always hidden in the back of our mind. Galileo, speaking of the possible existence of extraterrestrial life, said that it is not possible that beings similar to us exist in the Universe. On the contrary, he thought, it is quite likely that beings exist somewhere in space so different from us that we could never even imagine their aspect. These words are still valid; it suffices to watch any *Star Trek* episode to realize how terribly anthropomorphic are the aliens of science fiction. And this is not only because of the obvious difficulty of disguising an actor to play a non-anthropomorphic being; it is also because of the real difficulty of imagining such a creature. The basic problem is that a true general definition of living or intelligent being doesn't exist; we have just a single example of life and intelligence that we see every day.

One of the basic hypotheses of modern science is that physical laws are the same in the whole Universe and don't change in time; without this assumption we could not, for instance, interpret astronomic observations of distant galaxies, which are related to objects that existed millions or billions of light-years from us, in an equally remote past. But if we limit ourselves to objects in our galaxy, that hypothesis is a certainty. If that is so, however, are chemical and biological evolution determined by physical laws to the point that only one biochemistry, only one way for encoding genetic information, only one cellular structure is possible? Does life necessarily lead to eukaryote cells, multicellularity, differentiation of tissues, and so on? If so, extraterrestrials won't be very different from us, convergent evolution may lead to striking similarities. A being that evolved on a small planet, in a weak gravitational field, will of course be more slender, endowed with less powerful muscles, and, if the atmosphere is less dense, will have greater lungs, but it *will* always have muscles and lungs and—if the planet is illuminated by a star—eyes to see. And, since there are good reasons for intelligence to have evolved in a biped, with his eyes in a frontal position to allow binocular vision, intelligent aliens could well be very similar to us. This is clearly an extreme hypothesis, which can also be generalized to include psychological aspects of the nature of intelligent beings.

The opposite hypothesis is that there are many ways in which living beings can evolve and that life based on a biochemistry very different from ours may exist. Environments that are favorable to life can be very different, and every one could produce beings that have little in common with those evolved in different places. If this is the case, it could be very difficult to recognize a living being very different from us, and even more difficult to recognize such a being as intelligent. The very definition of intelligence could be difficult, or even impossible. And though in theory it may be easy enough to define what a conscious, sentient being is, in practice it might be impossible to communicate sufficiently with such a being to understand its existence.

Between these cases are a vast number of intermediate possibilities, and a correspondingly large number of hypotheses have been formulated. But they won't have any scientific validity until experimental verification is obtained. Indeed, we not only tend to think of extraterrestrial life or intelligence in human terms, we even refer to the present historical moment and to what is familiar to us: the hypothetical aliens end up thinking as humans of the planet Earth, possibly of a Western culture, living at the beginning of the twenty-first century. The technology we refer to is the present one, and this explains, for instance, the emphasis given to radio waves as the preferential means of communication in SETI efforts. This conditioning is perhaps even more deceitful than simple anthropomorphism: if we accept that it is absurd to expect extraterrestrials to have, for instance, two hands with five fingers each, we must also stop assuming that they use certain technologies or follow logical paths that seem natural to us.

Despite all these theoretical and practical difficulties, the search for extraterrestrial life continues, with theoretical studies, in-situ explorations performed by space probes bound to the nearby planets, and astronomic studies aimed at identifying habitable celestial bodies orbiting other stars.

The radioastronomical search for intelligent life has also made much progress, both in its theoretical elaboration and because the power of the instruments has increased enormously. But though some doubtful signals and some false alarms have been detected, no certain evidence of contact has yet been obtained.

At present humanity is on the eve of an important passage in its history. Humanity has just learned to move in space and is close to changing from a species living on a single planet to one spread in its planetary system, at first, and then perhaps in an even larger environment. Today, space is often seen as a laboratory, a place in which to perform scientific experiments and, increasingly, a place in which to develop economic activities (circumterrestrial space, at least). But real exploration and colonization projects await us. The scheduling of these developments is most uncertain, as shown by the failure of almost all forecasts of this sort in the past decades, and it is doubtful that our present civilization will succeed in exploiting the great opportunities offered by expansion in space. Nevertheless, even if our expansion in space should falter, if our species will not come to an abrupt end, our descendants will again start the trend of expansion in space that we were not able to pursue.

Other, perhaps more important, questions will thus be added to those regarding the existence of extraterrestrial life and of intelligent and conscious beings:

- Could humanity ever come in contact with these intelligent beings?
- Will it ever be possible for humans from Earth to participate, together with the other intelligent species that perhaps populate the Universe, in a larger community?
- Will each species, because of the enormous cosmic distances, be in complete isolation forever, even if the existence of many intelligent beings should become a certainty?
- Will humanity ever be able to entertain relationships with them more direct than those that radio links over very large distances can afford?

- Could it become possible one day to obtain information from a sort of cosmic *database* to which all species contribute information about themselves?

These are big questions: the knowledge that we are not alone in the Universe would have an enormous impact on our worldview, but it is the possibility of close contact and a mutual understanding with other intelligences and other civilizations that is really important and may have a huge influence on the future development of humanity.

1

THE HISTORICAL AND PHILOSOPHICAL PERSPECTIVES

THE MAGICAL VISION OF THE NONHUMAN

THE idea that humans are not alone in the Universe is ancient and lost in the mist of mythology. The myths of all ancient peoples are crowded with intelligent beings, often endowed with magical powers greater than those of humans. Often they dwell on Earth together with humankind or, if their abode is extraterrestrial, they live in a heaven that has little in common with the physical Universe. Actually, most ancient men and women could not distinguish between the physical

1

Universe and the spiritual world. Astronomy dealt with fantastic creatures of all the types and, even in relatively recent times, the *theological* space occupied by God and the angels in Christian tradition often coincided with *astronomical* space.

Animals, imaginary beings, and even natural phenomena were humanized and had human feelings, vices, and virtues. The gods of almost all ancient religions were very similar to men and women, even if their bodies could have the shapes of animals and their divine nature freed them from the limitations typical of the human condition, above all giving them the gift of immortality. Every human group had a myth of the origins, explaining how the world had come into existence and how humanity started, usually from a couple of ancestors. In many cases such myths gave the grounds to affirm the superiority of that group over all the others.

In this situation it made little sense to wonder whether humankind was alone in the Universe, since it was surrounded by gods, demons, and other intelligent beings, intellectually very similar to human beings, who shared the same physical world, while being able to transcend its limitations. Besides, the definition of *human* was usually only applied to the members of the same community.

If in ancient times the idea of intelligent beings coming from other celestial bodies was hardly conceivable, the myths concerning ancient gods were recently reinterpreted in this way, and many people are convinced that they have found the traces of extraterrestrial visits in ancient legends, sculptures, and drawings. In many of these scenarios, intelligent beings from the stars are alleged to have interacted with our ancestors, influencing the development of civilization on our planet. We will return to this subject in Chapter 5.

ANCIENT PHILOSOPHY

Almost all ancient peoples were good observers of the sky and succeeded in describing the apparent motion of the stars—thanks in part to far better conditions of visibility than those prevailing today, particularly in Europe—with fairly good precision. Nevertheless, their interpretation of what they saw was very far from what we have scientifically ascertained today. What they lacked was a scientific understanding that would have enabled them to make sense of the apparent motion of the stars, and to turn the bidimensional image of the celestial sphere into a coherent three-dimensional picture of the Universe surrounding us.

This scientific understanding of the world was exactly what the early Greek philosophers tried to develop. Thales (end of the seventh century BC) is usually considered the first important Greek philosopher and "natural scientist." He was deeply interested in astronomy and became popular with his countrymen by predicting the Sun eclipse of 585 BC, using the records of Chaldean astronomers with great skill. In general, Greek astronomy, heir to the Babylonian tradition, achieved remarkably good results in the prediction of eclipses and even formulated the concept of a spherical Earth, isolated in space.

The question of the shape of Earth was settled by Eratosthenes, who, at the end of third century BC, not only found strong evidence that it was spherical, but also succeeded in computing the diameter of the Earth from his own experimental observation with remarkable precision: his results differ from the actual value by less than 1 percent!

At the beginning of the fourth century BC, Eudoxos, a disciple of Plato, tried to explain the shape of the Solar System in a scientific way. He adopted a geocentric view, with Earth fixed in space at the center of the Universe and the Moon, the Sun, and the planets orbiting around it, being fixed on ideal spheres rotating around their poles. However, to explain the observational data, he had to make his model more complex by adding other spheres in order to obtain trajectories that were not circular. His disciples later had to add other spheres in order to obtain better agreement with observations, making the whole system even more complicated.

A few years later Heraclides Ponticus, another disciple of Plato, realized that the motion of some planets, namely Venus and Mercury, was better explained assuming that they rotated about the Sun and not Earth. He suggested a model in which Earth was still at the center of the Universe with the Sun, the Moon, and some planets orbiting it, but in which the other planets were orbiting the Sun.

At the end of the same fourth century BC, Aristarchus of Samos put forward a new idea: the Sun was at the center of the Universe, with all planet orbiting it and the Moon orbiting Earth. The two basic models, geocentric and heliocentric, were present at the same time, with a majority of philosophers supporting the first one. But almost all followers of both geocentric or heliocentric doctrines thought that beyond the Solar System there was a sphere of fixed stars, a rather vaguely described sphere rotating around Earth (in the geocentric model), on which were located a great number of bright points. The stars were therefore simply bright dots, all at the same distance from the center of the Universe.

However, different ideas were also present; Anaximenes, for instance, held in the sixth century BC that the stars were made of fire like the Sun, with planetary bodies made of earth and water orbiting around them, but not observable from Earth. In the following century, Democritus developed a new view of the Universe, in which all matter was made up of microscopic constituents, which he termed *atoms*, moving without rest in an infinite empty space. In this Universe, infinite both in space and time, there was an infinity of worlds.

The two concepts, that of a universe enclosed by the sphere of the fixed stars containing only the Sun and the planets, and that of an infinite universe with innumerable suns and planets, coexisted. Over time, for those who believed that the planets of the Solar System, the stars, and the hypothetical planets orbiting around them were actual celestial bodies and not simply bright spots on the celestial sphere, the problem of their habitability and of the presence of extraterrestrial beings, perhaps intelligent ones, started to become a subject of discussion.

Aristotle, in his effort to systematize the scientific knowledge of his times, embraced the geocentric view of the cosmos, with a universe limited by the sphere of the fixed stars. Besides, he thought that celestial objects were made of a substance different from that constituting everything that can be found below the sky of the Moon; while Earth is made up of the four elements earth, water, air, and fire, celestial bodies are made of a perfect substance, the quintessence, or "fifth element," or ether. Everything that exists above the sky of the Moon he thought to be perfect and to move in circular orbits. The stains clearly visible on the Moon were attributed by his followers to impurities emanating from Earth or simply to reflections of the lands and seas of our planet. In a view of this kind it was absolutely unthinkable that the stars could be inhabited, other than by divine beings or other perfect creatures.

But the physics of Aristotle never gained universal acceptance in ancient Greece, and other theoretical formulations had many followers. Some Greek philosophers did not believe in a clear-cut distinction between the sublunar and the celestial world.

Thales, for instance, held that the Moon was a spherical body, with a nature similar to that of Earth. Such ideas gave rise to fantastic stories about beings living on other worlds, sometimes endowed with powers far greater than those of men, but nevertheless not divine. The followers of Pythagoras (sixth century BC) thought that the Moon was inhabited by huge plants and animals—a thesis resumed centuries later by Plutarch (second century AD) in his essay *De Facie Orbe Lunae*. These ideas brought some writers to imagine trips on other worlds; the most famous work is

The True History, written by Lucianus of Samosata around 177 AD. The philosophical school that went furthest in this direction was that of the Epicureans, supporters of the atomistic physics of Democritus. In his *De Rerum Natura*, Lucretius (first century BC) wrote:

> My mind asks explanations, since the Universe is infinite. What is there then out there, beyond the boundaries of the world, as far as where the mind cares to look and where the rush of thought freely flies alone. . . . For in no way can it be considered likely—since that space is endless, extending in every direction, and the atoms are infinite in number and fly in enormous quantity in many ways, pushed by an eternal movement—that only this world and sky have been created and so many atoms do not do anything outside; particularly since this same world was made by nature when the atoms, spontaneously colliding at random, united in many ways by chance, at random and in vain they finally joined to generate great things: the earth, the sea, the sky and the breed of animate beings.
>
> It is therefore more and more necessary that you admit that elsewhere other aggregates of matter exist, in the same way as this which is contained by the air with an ardent embrace . . . It is necessary to admit that other globes exist in the space, other breeds of men and races of animals . . . It has to be admitted that the sky, earth, sun, moon, sea and all the other things that exist, are not unique, but in innumerable number . . .

In their conception of an infinite universe in which atoms gathered to generate all material things, an infinity of suns, planets, and living beings of every type can exist. In this formulation the plurality of inhabited worlds is a necessary consequence. Metrodorus, one of the exponents of the Athenian Epicurean school, wrote: "To consider the Earth the only populated world in infinite space is as absurd as to assert that in an entire field of millet, only one grain will grow."

In Greek philosophy the term *plurality of the worlds* often didn't mean plurality of planets, perhaps habitable, in our Universe but plurality of universes, each one limited by its own sphere of fixed stars, not observable by us but nevertheless existing. In a completely different context, this idea has recently been reformulated by some scientists, who replace the classical Universe with a multiverse. The term *plurality of the worlds* only began to mean univocally *plurality of planets* with Galileo. Aristotle could not accept this plurality of universes, since within the frame of his philosophy it would implicate a plurality of natural places and a plurality of prime movers—clearly unthinkable concepts.

Both the geocentric and the heliocentric systems had serious problem explaining the observational data, owing to the fact that orbits, as a consequence of the existence of *celestial spheres* to which the planets were attached, were assumed to be circular. In Aristotelian physics orbits had to be circular also because of the perfection of the celestial world, in which no trajectory less perfect than a circle could be imagined. Yet even philosophers who tried to explain nature without resorting to metaphysical ideas could not imagine anything different. Supporters of the geocentric theory imagined a complicated sequence of spheres, one rotating around the other, to explain with increasing precision the motion of planets as observed. Finally the great astronomer Hipparchus (second century BC) systematized the geocentric system to account for observational data by describing a whole set of complex motions (epicycles, deferents, etc.) that were superimposed on the simple circular orbits along which the celestial bodies moved in their diurnal motion around Earth. In the first half of the second century AD his very complicated system was eventually enshrined by Claudius Ptolemaeus, who gave it its final form. The geocentric system, which was fully compatible with the Aristotelian doctrine and above all explained astronomical observations far better than heliocentric systems, became the standard paradigm for the interpretation of the world.

MEDIEVAL PHILOSOPHY

In late Roman times and during the early Middle Ages, philosophy was much concerned with metaphysics and ethical problems. The study of nature, that is, science as we now intend it, languished. The few who were interested in astronomy accepted the Ptolemaic system almost unquestioningly. When, at the beginning of the second millennium, European scholars started again to deal with *natural philosophy*, the original texts of Greek philosophers were mostly lost, and much of their work was dedicated to restoring ancient knowledge. For example, Johannes Scotus Eurigena, who from 847 to 870 AD headed the Palatine School founded by Charlemagne, denied the division of the Universe in an earthly and a celestial world and supported the mixed (geocentric and heliocentric) system devised by Heraclides Ponticus.

Aristotelian philosophy was mainly known to medieval European scholars through the commentary of Arab thinkers, first among them that of Averroës (Ibn Roschd), whose interpretation of Aristotle was mainly incompatible with Christian doctrine. Although at first the Christian Church didn't support the Aristotelian worldview, its hostility to Epicurean philosophy was even greater, and certainly played an important role in Catholic opposition to the idea of the existence of other habitable worlds.

Slowly, however, attempts to interpret the Aristotelian view of the world in a way that would be compatible with Christian theology were made. It was Saint Thomas, in the middle of the thirteenth century, who finally succeeded in this task. The new interest in Aristotle's physics was particular useful in promoting studies of the physical world, which were absent in the works of most earlier medieval scholars. This was without doubt an important step in the direction of a new kind of science, but unfortunately it was much weighted down by literal interpretations, which did not take into account the new ideas proposed by Alexandrine and Arab philosophers.

The role played by Saint Thomas in the Christian world was performed somewhat earlier by Maimonides (born in Cordoba in 1135), in Jewish culture. In his *Guide for the Perplexed*, he tried to harmonize Aristotelian philosophy and Jewish theology.

At the end of the thirteenth century Dante Alighieri, who faithfully followed the Aristotelian description of the Universe, not only believed

FIGURE 1.1 *Medieval cosmological representation, in its simplest version. The Earth is flat and is covered by the hemispheric celestial dome. In the print a traveler reaches the limit of the Earth and, poking his head through the celestial dome, observes the mechanism that moves the stars.*

that no other habitable worlds exist, or better, that no other world exists at all, but also that humans live only in one hemisphere. Ulysses, when exhorting his companions to pass beyond the Pillars of Hercules, admonished them not to deny themselves the knowledge "... di retro al sol, del mondo sanza gente" (following the Sun, of the world without people).

But if for Dante and for many medieval scientists Earth was spherical, the idea of a flat Earth with the celestial dome, reduced to a semisphere (Figure 1.1), over it, gained new momentum in the Middle Ages. But Aristotelian physics continued to raise not a few doubts, particularly in ecclesiastic circles, so the new interest in physics ended up giving the idea of an infinite universe the chance to gain ground again.

In 1277 Pope John XXI asked Stephen Tempier, bishop of Paris, to condemn the Averroistic, and therefore essentially Aristotelian (Latin Averroism is often referred to as radical Aristotelism), tendencies that were spreading in the University of Paris. Tempier published 271 "sentences" that had to be accepted by all believers. Sentence 34 states that whoever believes that God cannot create a number of worlds is a heretic; the plurality of worlds is not expressly affirmed, but to deny it, at least as a possibility, is heresy. The same pope was the author of many scientific and philosophical texts, above all in the field of logic, in which he tried to free himself from Aristotelian orthodoxy and to create a new science.

In the same period also in the Jewish cultural environment there was a reaction against Aristotelism in the form elaborated by Maimonides.

THE RENAISSANCE

The fifteenth century saw an increasing interest in the physical world and a deeper study of the scientific works of Greek and Alexandrine philosophers. Renaissance thinkers felt freer to study the physical world in a way that was independent from religion and theology, pushing this separation to the point of admitting the existence of a double truth. This led on one side to rediscovering the atomistic view and, more generally, epicurean philosophy, but on the other side it strengthened the Aristotelian interpretation of the world, without bothering too much with the conflict between Aristotle and the Christian worldview, which still remained.

One of the better known supporters of the plurality of worlds was Cardinal Nicola Cusano, who, in his essay *De Docta Ignorantia*, published in the middle of the fifteenth century, affirmed the existence of endless worlds, probably inhabited, in orbit around other stars similar to the Sun. His cosmology is based on philosophical arguments, and one of his main proofs of the infinity of the Universe was the fact that it had to mirror divine infinity. Nevertheless, some modern ideas can be found in the works of Cusano, such as using mathematics and experimental science as a basis for philosophy. Cusano found a copy of Pliny the Elder's *Natural History* and a large number of other manuscripts. In the fifteenth and sixteenth centuries, humanists found a large number of classical texts, which were believed to have been lost; this caused a rediscovery of classical philosophy, including that of the Epicurean school, and led to increased questioning of the Aristotelian orthodoxy. In 1488 Lorenzo Valla translated the writings of Aristarchus of Samos, giving new life to the heliocentric doctrine.

Giordano Bruno, who often took inspiration from Lucretius and from atomism, was a champion of the plurality of inhabited worlds and wrote an essay, *De l'Infinito Universo et Mondi* (On the Infinite Universe and Worlds), in which he affirmed that the stars are celestial bodies similar to the Sun.

He denied both the geocentric and heliocentric systems, suggesting that actually the Universe doesn't have a center. The reasoning that brought him to defend such a position was still, however, theological (to deny the infinity of the Universe and the plurality of worlds amounts to denying the infinite power of God) and not scientific; the astronomical observations of the time were insufficient either to confirm or to refute statements of this kind. His view of the world was still animistic and magical: Bruno, despite his great appeal to later philosophers, is much less modern than Copernicus and other great astronomers of his time.

Owing to his disagreement first with the Catholic Church and then with the Calvinist Church, he moved from Italy to Switzerland and then to France and England, where he was safe from accusations of heresy. However, he later returned to the Continent, first to France, then to Germany, and finally to Italy. He was arrested in Venice in 1592 and brought to Rome the following year, where he was subjected to a trial by the Holy Inquisition, which lasted seven years and ended with his condemnation. Even if one of the thirty charges that caused him to be burned at the stake in 1600 was his belief in the existence of innumerable worlds, it was a far less serious charge than some others of more direct theological relevance, like denying transubstantiation. To assert that Giordano Bruno's death sentence was caused by his scientific and astronomic beliefs is wrong, particularly because none of the other philosophers who defended similar theories suffered such an inhumane sentence.

Chapter 1

THE BIRTH OF MODERN SCIENCE

The publication in 1543 of Copernicus's essay *De Rivolutionibus Orbium Coelestium* (The Orbits of Celestial Bodies) was the start of an astronomical revolution that radically changed the cosmological perspective but, in comparison with the physics of Aristotle, did not essentially change the extension of the Universe. The Copernican Universe is still a finite Universe, limited by the crystalline spheres, even if it seems that he had some doubts about the sphere of the fixed stars; in Chapter 8, Book 1, of the *De Rivolutionibus Orbium Coelestium*, he does speak of an "endless sky."

Copernicus had studied in the universities of Padua and Bologna, where the writings of Aristarchus, translated by Valla, were discussed. His teacher in Bologna, Domenico Maria da Novara, was probably convinced of the validity of the heliocentric theory. But if the heliocentric hypothesis was not new, Copernicus reassessed it and set it on more solid scientific ground, confronting the theory with experimental observations.

He defended its superiority in comparisons with the Ptolemaic theory on the grounds of a greater perfection and simplicity, since the latter had to resort to a complex system of epicycles and deferents to agree with observations. Nevertheless, seeking the maximum perfection and simplicity, Copernicus also assumed that the orbits of the planets were circular. For this reason his system did not appear to be in accord with astronomical observations any more than that of Ptolemy, and many astronomers were not convinced by it.

Tycho Brahe, the last great European astronomer who performed accurate observations without optical instruments, ended up introducing an intermediate system, in which the Moon and the Sun rotated around Earth, while the planets rotated around the Sun.[1] However, he realized that the celestial world is not unchangeable when he observed, in the constellation of Cassiopeia, a "new" star (*stella nova*, according to his definition, is still in use today for stars undergoing a sudden explosion).

In 1577 the English mathematician Thomas Dix wrote that the stars were uniformly distributed in the infinity of space. Finally, there was no longer a sphere with fixed stars, all at the same distance, but a true three-dimensional distribution of stars.

In 1608 the first telescope was presented at the Frankfurt fair. Galileo immediately realized the importance of the new instrument for astronomy and, after building an improved version, began his observations. In a few

---

[1] This system is equivalent to the Copernican system with a simple change of reference frame. The systems are not distinguishable using only astronomical observations.

months he made a series of discoveries that definitely proved the distinction between the sublunar world (imperfect) and the celestial world (divinely perfect) to be false: the surface of the Moon was not perfectly spherical, but had mountains and valleys; the Sun had black spots; the planets were not bright dots, as the stars, but disks; and Venus had phases like the Moon. The *Sidereus Nuncius*, published in 1610, put in place the foundations of modern astronomy.

In his *Istoria e Dimostrazioni Intorno alle Macchie Solari e Loro Accidenti* (History and Demonstrations on Sunspots and Their Details, 1613), Galileo states that he "considered as a false and condemnable point of view to assume the existence of inhabitants on Jupiter, Venus, Saturn and the Moon, intending for 'inhabitants' animals like ours, and particularly men." But then he goes on to write that "it is possible to believe that living beings and plants exist on the Moon and the planets, whose characteristics are not only different from those of beings on the Earth, but also from what our wildest imagination can produce."

These words of Galileo's contains perhaps the first explicit position against an anthropomorphic view of possible extraterrestrial beings, a warning that is undoubtedly still important today.

While Galileo completed his astronomical observations, Johannes Kepler, using the results of Tycho Brahe's observations, realized that the orbits of the planets were not circular but elliptical. This settled the controversy between the heliocentric and the geocentric system once and for all, since in this way Kepler was able to explain the observations with very good precision. Moreover, it was a final blow to the idea of the substantial difference between the sublunar and the celestial world. The main obstacle against the existence of other *breeds of men*, as Lucretius had called them, was removed. Less than one century later, Newton, with his gravitational theory, which gave a theoretical basis to the empirical rules discovered by Kepler, showed that the laws governing the motion of celestial bodies are the same as those that apply to the motion of objects on Earth.

Kepler, in a text published in the form of a letter to Galileo, spoke of the possibility of space journeys to the Moon and planets, and didn't oppose the idea of their habitability. Rather, he stated that it was likely that inhabitants of Jupiter and Saturn existed and, years later, wrote a novel based on a trip to the Moon.

With detailed knowledge of the Solar System and the hypothesis that the stars were no more than very distant suns, the problem of the existence of extraterrestrial life took on two distinct facets: the habitability of the planets in the Solar System and the existence of other planetary systems, perhaps with habitable—or, indeed, inhabited—planets. In 1647 the astronomer Johannes Hevelius published an important scientific work entitled *Selenografia*, in which he stated that the air on the Moon must be extremely thin and cast doubts on the existence of water there. The habitability of the Moon was therefore strongly questionable.

In the first half of the seventeenth century, René Descartes developed a theory explaining the formation of the Sun and the stars at the center of vortices of particles of different kinds, some of which would have formed planetary systems. A consequence of such a model is the existence of planets around other stars. Even if Descartes did not say anything on the possible existence of intelligent beings on such planets—and the theory of the vortexes was later definitely disproved by Newton—Cartesian cosmology brought other scientists toward the acceptance of the plurality of inhabited worlds.

The book *Conversations on the Plurality of Worlds*, published by Bernard Le Bovier de Fontenelle in 1686, one year after Newton's *Principia*, was a huge success, with thirty-three editions in French and many translations published before the death of the author. It won its author the election to the Academy of Sciences in Paris and accustomed a large public of average culture to the Copernican system and to the idea that many inhabited planets could exist in the Solar System.

Huyghens (1629–1695) tried for the first time to measure the distance of the stars, calculating that Sirius was about half a light-year from the Sun. Even if the numerical value was much smaller than the true one (Sirius is actually 8.7 light-years away), he correctly concluded that it was not possible to observe planets in orbit around other stars at such a distance and therefore the absence of direct observational evidence did not mean that they do not exist. He thought that not only did extrasolar planets host life, but also that intelligent beings lived on them, with their own science, art, and philosophy.

In 1725 the reverend William Derham published a book in which he tried to reconcile theology with the new vision of the Universe. He stated that not only the planets but also the Sun and the comets were inhabited. The book was a success and was reprinted several times.

A number of theories on the origin of the Solar System were put forward in the eighteenth century. They can be grouped into two antithetical types: The first, which had Descartes as a forerunner and Kant (1775) and Laplace (1796) as its best known exponents, held that the Solar System was formed by the same evolutionary process that gave birth to the Sun and the other stars. According to this group of theories a large number of stars (perhaps all) should have a planetary system, and the number of extrasolar planets is very large. The second group of theories explain the origin of the Solar System with a catastrophic event, like the passage of another star close to the Sun or the explosion of a star orbiting around the Sun (which clearly was assumed to be originally a double star). Since such catastrophic events are extremely rare, the majority of the stars, perhaps all of them, have no planets, and the formation of the Solar System must be regarded as an extremely rare, perhaps unique, event.

In the past two centuries these two theoretical formulations fought each other, but neither prevailed. Then, recently, the discoveries of extrasolar planets and of planetary systems in formation seemed to definitely settle the question in favor of the first model. In the nineteenth century the idea that many, perhaps all, the planets of the Solar System were inhabited was almost taken for granted. At the beginning of the century, the reverend Thomas Dick wrote ten books on the subject and, for the first time, even tried to calculate the population of the various planets. With computations based on the population of England in his time, he obtained the results shown in Table 1.1.[2] It should be noted that

TABLE 1.1 *Population of the various planets, asteroids, and satellites according to Thomas Dick (1837)*

Celestial body	Population
Mercury	8,960,000,000
Venus	53,500,000,000
Mars	15,500,000,000
Vesta	64,000,000
Juno	1,786,000,000
Ceres	2,319,962,000
Pallas	4,000,000,000
Jupiter	6,967,520,000,000
Saturn	5,488,000,000,000
Rings of Saturn	8,141,963,826,080
Uranus	1,077,568,800,000
Moon	4,200,000,000
Satellites of Jupiter	26,673,000,000
Satellites of Saturn	55,417,824,000
Satellites of Uranus	47,500,992,000
Total Solar System	21,891,974,404,080

[2] The table is taken from G. Lemarchand, *El Llamado del las Estrellas*, Lugar Cientifico, 1991.

in so doing he neglected the possible presence of oceans and obtained unbelievable results even for Earth, which, extrapolating the method used by Dick, should have more than 55 billion inhabitants.

Finally, in 1838, the measurement of the parallax of the nearest stars, and therefore of their distance, made it possible to calculate the intrinsic brightness of stars, thus showing that they are celestial bodies similar to the Sun. This was the first experimental verification of an idea that had been around for centuries but had, until then, been just a hypothesis.

Sir John Herschel, son of William Herschel, musician and famous astronomer who had made a number of discoveries—such as those of the planet Uranus and of two satellites of Saturn—was himself an astronomer. To complete, with observations of the southern sky, the catalogue of stars begun by his father, he founded the observatory of the Cape of Good Hope, where he spent many years. While he was in South Africa, a reporter of the *New York Sun*, Richard Locke, published a series of articles beginning on August 25, 1835, in which his discoveries were described.

The reporter began describing the astronomer's powerful telescope, able to show objects only 45 centimeters long on the Moon, and continued the escalation day after day with further revelations, building to a delirious climax. The readers were informed about the discovery of flowers and plants, then of animals, and finally of intelligent beings, of which the reporter even gave sketches (Figure 1.2). This was probably the first in a long series of mystifications based on extraterrestrials, and, as were many of those that would follow, it was a runaway success.

The fact that the August 26th issue of the *New York Sun* sold 19,000 copies, an absolute record for the time, shows both the public's interest in the subject and its gullibility. Herschel didn't have any role in the fraud, as his colleagues immediately recognized; to them, the absurdity of claiming that a telescope could have such power—not just orders of magnitude greater than the instruments of the 1830s, but even of the most powerful of today's instruments—was evident. Actually, the articles were meant by Locke to be a satirical attack against the many philosophers and scientists who at that time were supporting the idea of the plurality of the worlds, but ironically it misfired: the desire of the public to believe in it was such that it was taken seriously.

The number of followers of the idea that many worlds were inhabited by intelligent beings at the beginning of the nineteenth century is also attested to by the fact that the founders of the two Christian denominations that appeared at that time in the United States—the Seventh Day Adventist Church and the Church of Jesus Christ of Latter-Day

(a)

(b)

FIGURE 1.2 *The Great Moon hoax of 1835. (a) Sketch of an inhabitant of the Moon allegedly seen by Sir John Herschel with his telescope, from the frontispiece of the book* Delle Scoperte Fatte Nella Luna dal Dottor Giovanni Herschel *(On the Discoveries Made on the Moon by Dr. John Herschel), published in Naples in 1836. (b) Illustration from an 1836 English pamphlet based on the descriptions published in the* New York Sun.

Saints—believed in it to the point of including the plurality of the worlds in their writings.

FIRST ATTEMPTS AT CONTACT

In this climate of almost complete certainty that the Solar System was inhabited, the first attempts at contact began. The first proposal was put forward by Karl Friedrich Gauss in 1820; the mathematician suggested that a large quantity of trees be planted in Siberia, creating a drawing on the ground in the form of a right-angled triangle with squares built on the sides and on the hypotenuse. The hypothetical inhabitants of the Moon, looking at the figure with powerful telescopes, would conclude that intelligent beings lived on Earth. The idea that mathematics constitutes an universal language, understandable by any intelligent being, is still widespread today, and Gauss's suggestion is not that different, other than in the form of the message, from recent ideas and attempts. The Gauss proposal was not recorded by the great mathematician in any of his writings, but was reported by journalists of the time. In a letter he wrote in 1822 to astronomer Wilhelm Olbers, he also suggested using 100 mirrors, each with a surface of one square meter, to communicate with the inhabitants of our satellite.

On the same lines, the director of the astronomical observatory in Vienna, Johann von Littrov, suggested in 1840 that a huge trench of circular or square shape be dug in the Sahara desert and filled with oil. Once set on fire, the line of flames with a distinct geometrical shape would certainly have shown the existence of intelligent beings on the Earth. A few decades later, in 1879, Charles Cros, proposed to use large mirrors scattered in Europe or electric lamps with mirrors to focus the beam of light to communicate with the inhabitants of Venus and Mars. Despite the complexity of the language he suggested, based on flashes of light of different colors to produce different figures, the basic idea was always that of sending optical messages.

The developments in the field of electromagnetism, the discovery of electromagnetic waves by Hertz, and their subsequent application by Marconi supplied a new tool for communicating with extraterrestrials. In 1899 Nikola Tesla—the brilliant Croatian engineer who replaced Edison's direct current grid with alternate current distribution—built an imposing plant in Colorado Springs. It consisted of a coil with a diameter of 23 meters connected to a 60-meter-high antenna and he actually used it to broadcast signals toward space, the first messages intentionally sent by humanity toward the cosmos. His equipment was also used as a detector and did receive strange, regular noises. Today we know these to be a natural phenomenon due to electrical discharges in the high atmosphere, but at the time the phenomenon was unknown and Tesla actually thought he had received signals from an alien civilization.

In 1922 Guglielmo Marconi also made attempts to receive extra-terrestrial signals on his yacht *Elettra*—in the middle of the Atlantic Ocean, to avoid interference—and he received the same signals, but was also unable to realize that they were of natural origin.

COSMISM

Between the end of the nineteenth and the beginning of the twentieth centuries a philosophical doctrine, which was given the name of *Cosmism*, flourished in Russia. Its main exponents were the philosopher Fedorov and the teacher Kostantin Tsiolkovsky, but Cosmism influenced many important intellectuals of the time, like Dostoevsky and Tolstoy.[3] Tsiolkovsky is known more for his contributions to astronautics than for his philosophical ideas or for his views on life in the Universe and extraterrestrial intelligence. While his research in astronautics and even his work as a consultant for science-fiction movie productions were broadly publicized by the Soviet regime, his philosophical works, which were in contrast with the official materialism of the Soviet Union, were only recently made known to the public. For the people of the Soviet Union, Tsiolkovsky was a hero, the pioneer of astronautics celebrated by postage stamps, propagandistic brochures, and statues in the style of socialist realism.

Cosmism preached the unity of humanity (or better, of the innumerable communities of intelligent beings that, according to Tsiolkovsky, evolved in the cosmos) and the Universe, and attributed to humanity a cosmic destiny. It was a sort of mixture of spiritualistic philosophy and admiration for modern science and technology, with the latter seen as instruments for the human spirit to transcend the narrow limits of the earthly environment and launch itself into space adventure, creating bonds of brotherhood with all other intelligent

---

[3] V. Lytkin, B. Finney, and L. Alepko, "Tsiolkovsky, Russian Cosmism and Extraterrestrial Intelligence," *Q.J.R. Astr. Soc.*, vol. 36, pp. 369–376, 1995.

beings in the cosmos. But, as we will see in Chapter 4, the *space imperative* that pushes humanity to abandon its cradle, as Tsiolkovsky put it, to start a new life in the vast Universe, and the existence of other intelligences, at least within reasonable distances (on a galactic scale), *can* be in conflict.

If the *space imperative* is a characteristic of all intelligent species, and if extraterrestrial intelligences exist not too far from our Solar System, then they should have revealed themselves to us a long time ago. This problem is often referred to as the *Fermi paradox* (or *Fermi question*), but it was actually forwarded by Tsiolkovsky many years before Fermi. This question will be discussed in detail in Chapter 5.

FROM ENTHUSIASM TO DISENCHANTMENT

The absurdity of the stories filed by Richard Locke and the fact that astronomers doubted the presence of an atmosphere and water on the Moon did not shake the faith of the public, or even the opinion of scientists; at most they caused a shift of interest from the Moon to other celestial bodies. If the Moon didn't host living beings, surely Mars, a twin planet of Earth, was inhabited: the stories about the dwellers of the Moon left the stage to those on the Martians, which had a real boom at the end of the nineteenth century.

The three main players were three astronomers, the French Camille Flammarion, the Italian Giovanni Schiaparelli, and the American Percival Lowell. Schiaparelli, appointed in 1862 as director of the Brera observatory in Milan, began his observations of Mars during the 1877 opposition, when Mars was at the shortest distance from Earth and under the best light conditions for detailed observation (Mars is in opposition when it, Earth, and the Sun are on the same straight line, with the two planets on the same side of the star, about once every two years). His observations went on for more than 20 years, but he only published the results of those performed in the oppositions from 1877 to 1890, since he feared that his weakening sight might later prevent him from obtaining precise results. This concern is proof of his scientific and intellectual correctness.

He traced a large number of maps of the surface of the planet, noting the polar caps that changed in extension according to the seasons, and observing in detail the darker zones, which he interpreted as seas, and the brighter ones, which he obviously took as continental masses. In the latter

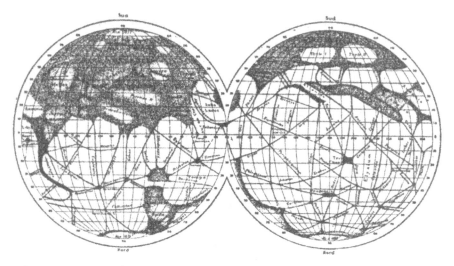

FIGURE 1.3 *Map of the two hemispheres of the planet Mars drawn by Schiaparelli following his observations performed since 1877.*

ones he observed periodic variations of color, which he ascribed to vegetation cycles. But his most important observations were those related to the so-called *canali*, a net of dark lines that practically interested all the brighter zones of the planet (Figure 1.3). In his articles he made detailed descriptions of what he observed, avoiding interpretations and above all not making hypotheses on the presence of inhabitants of the planet; nevertheless the Italian term *canali* (meaning both natural and artificial waterways) was translated into English as canals, that is, artificial waterways, and this gave rise to the most fanciful interpretations.

But it would be wrong to affirm that Schiaparelli always abstained from interpreting his discoveries; while in scientific articles he was always very conservative, in a series of popular articles for the lay public he gave free rein to his imagination (he mounted the hippogriff, as he put it) and expressly noted that he thought the dark lines were actually zones of luxuriant vegetation surrounding artificial canals, built by an ancient civilization in an attempt to bring water from the melting polar caps toward the tropical and equatorial regions, and so survive the progressive desertification of the planet. The controversial phenomenon of the *gemination*, that is the splitting of almost all the canals in some seasons of the Martian year into two parallel lines, that had been observed by several astronomers, was explained with a complex system of dikes that alternatively flooded various zones along the sides of the canals to maximize crops. In a style always concerned with the likelihood of the hypotheses, and, not without a sense of humor, Schiaparelli extrapolated from the astronomical observations some theses on the history, sociology, and even political system of the inhabitants of Mars:

> It will be interesting to see what form of social system is more convenient to such a state of things, such as we have described; if the connections, rather the community of interests, that ties together the inhabitants of every valley, will perhaps make here the practice of collective socialism much more practical and suitable than on the Earth, making of every valley and of its inhabitants something similar to a huge phalanstery, in such a way that Mars could also become the heaven of the socialists.[4]

As if all that were not enough, the first spectroscopic observations of the planet seemed to confirm the presence of water vapor and oxygen in its atmosphere, and this further strengthened the certainty of the existence of forms of life.

Lowell was so thrilled by the descriptions of the Red Planet that were given by astronomers and scientific journalists that he became an astronomer and invested his huge patrimony in the construction of a great observatory at Flagstaff in Arizona, devoted to the study of Mars. Despite his attachment to theories that were not demonstrable at that time and were later proved to be completely wrong, he was for decades the highest authority on Mars and exerted a great influence on planetary astronomy.

In the last decade of the century the certainty of the existence of intelligent beings with an advanced technology on Mars was such that when, in the year 1900, a French foundation instituted the 100,000-franc Guzman prize for the first man on Earth who succeeded in establishing contact with an extraterrestrial, but contact with a Martian was expressly excluded. The jury clearly thought that such an enterprise was so easy as not to deserve any prize!

With the constant improvement of astronomical instruments, it was soon clear that the canals were an optical illusion; the human eye, when it tries to see details at the extreme limits of visibility, tends to connect, in straight lines, points that are scattered in a random way. Observed with better telescopes, the canals simply disappeared. Another disappointment came with the first reliable spectroscopic studies, which led to the conclusion that the Martian atmosphere did not contain meaningful quantities of water and oxygen.

Slowly, Martians left the scientific literature until they survived only in

---

[4] G. Schiaparelli, "La vita sul pianeta Marte," extracted from issue no. 11, year IV, 1895, of the journal *Natura ed Arte*, reported in P. Tucci et al. (editors), *Giovanni Virgilio Schiaparelli, La Vita sul Pianeta Marte*, Mimesis, Milano, 1998.

FIGURE 1.4 Hypothetical forms of Martian life inspired a drawing by Douglas Chaffee for an article by Carl Sagan in National Geographic *in 1965.*

science fiction. From *The War of the Worlds* by H.G. Wells to *Under the Moons of Mars* by Edgar Rice Burroughs, and from *Out of the Silent Planet* by C.S. Lewis to the *Martian Chronicles* by Ray Bradbury—to cite only the most famous titles—the inhabitants of Mars were the protagonists of a large number of novels and stories, but scientists had realized that the existence of intelligent beings on planets of the Solar System was in fact very unlikely.

Nevertheless, many scientists still thought that Mars and Venus hosted higher forms of vegetal and animal life, even if not so evolved as to be intelligent. Carl Sagan, for instance, in a 1965 article published in *National Geographic*, put forward the idea that the lack of an ozone layer on Mars would have compelled the various forms of life to develop a protective layer against radiation from the Sun (Figure 1.4).

However, neither the composition of the atmosphere nor the value of atmospheric pressure on the ground were known in detail and therefore there was no reason to exclude the possibility of the existence of some form of complex life.

In the same way, it was known that the surface of Venus was hidden from observation from space by thick clouds, which were thought to be water vapor. The fact that the planet was closer to the Sun than the Earth allowed some scientists to think that it was much hotter, and that the clouds could hide a surface with seas, swamps, and jungles, populated by big insects. The Nobel laureate Arrhenius, for instance, wrote in 1918:

> We must therefore conclude that everything on Venus is dripping wet. ... A very great part of the surface of Venus is no doubt covered with swamps. ... The temperature on Venus is not so high as to prevent a luxuriant vegetation. ... Only lower forms of life are ... represented, mostly no doubt belonging to the vegetable kingdom; and the organisms are nearly the same all over the planet. The vegetative processes are greatly accelerated by the high temperatures.

Starting from the end of the 1960s, the first probes approached Mars and Venus, resulting in a dreadful disappointment. The Russian *Venera* probes and the American *Pioneer* and *Magellan* probes found a much different state of affairs: the temperature on the surface is as high as 460°C, the clouds are made of droplets of sulfuric acid, the atmosphere is rich in carbon dioxide and nitrogen, and the pressure is crushing, about ninety times atmospheric pressure on the surface of Earth. The environment on Venus is therefore unsuitable for any form of life, at least as we know it.

Similarly, *Mariner 4*, the first probe that, after several failed attempts, performed a flyby of Mars in 1966, sent some extremely disappointing images back to Earth: Mars had a desolate aspect, very similar to that of the Moon. The following missions mitigated this first impression only in part: *Mariner 9*, which achieved an orbit around the Red Planet and continued to send images for almost a year, and above all the *Viking* probes, found traces of a Martian past in which liquid water flowed on the surface and the landscape was much less desolate and less similar to that of the Moon.

Nevertheless, the low pressure and temperatures and the composition of the atmosphere, together with the results of experiments aimed at searching for life, definitely canceled all hopes of finding higher living beings on Mars and, for the majority of scientists, even of finding bacteria or other very simple forms of life.

The very disappointing results obtained by the probes of the 1960s and 1970s brought many to pessimistic conclusions about the possibility not only of finding life in the Solar System, but also of exploring the planets with manned missions and then colonizing them.

Currently there is little hope of finding complex life-forms, not to speak of intelligent life, in the Solar System, and even the discovery of bacteria would be considered a very important result. Nevertheless, as will be seen in Chapter 3, in the last few years these pessimistic conclusions have been somewhat revised and some as yet timid hopes have been expressed.

2

THE RELIGIOUS
PERSPECTIVE

IS EXTRATERRESTRIAL LIFE A THREAT
TO RELIGION?

ARTHUR C. Clarke wrote in 1951 in *The Exploration of Space*
that there are people who are afraid that "the crossing of space,
and above all the contact with intelligent but non-human races,
may destroy the foundations of their religious faith." He then went on to
note that "in any event their attitude is one that does not bear logical
examination—for a faith that cannot survive collision with the truth is not
worth many regrets."[1] Actually the idea that the discovery of extraterrestrial life, and above all intelligence, will give such a blow to religion
to put all churches and religious institutions definitively "out of business,"

[1] Arthur C. Clarke, *The Exploration of Space,* Harper & Brothers, New York, 1951.

as they say, is fairly common among atheist scientists. This idea is not new: in the era in which the plurality of the worlds was accepted by the majority of people (religious and not), Thomas Paine wrote in *The Age of Reason* (1794): "He who thinks he believes in both [the Christian view of the world and extraterrestrials] has thought but little of either." And he would certainly choose to believe in extraterrestrials. Many SETI scientists would agree. Jill Tarter, for example, wrote: "God is our invention. . . . If we get a message and it's secular in nature, I think that says that they have no organized religion—that they have outgrown it." And this will cause us to do the same.

In a strange way the idea that a proof of the existence of extraterrestrial intelligence would put an end to religion is shared by the most conservative fundamentalist Christians. They are so sure that extraterrestrial intelligences do not exist that they think that any message from space, or even any evidence of extraterrestrial life, must be a fake fabricated by demonic creatures in their struggle against the faithful.

In many cases the belief in extraterrestrial intelligence takes on the character of a religion of its own, in competition with traditional religions, in some instances in an explicit way, as in the cults in which the faithful are engaged in a sort of messianic expectation for extraterrestrials to save them, but more often in a subtle and implicit way. The certainty of some scientists that advanced extraterrestrials will be benevolent and that contact with them will dramatically improve the human condition certainly has messianic overtones.

Interestingly enough, the idea that contact will put an end to religion is not shared by either religious people or representatives of the churches, with the already-mentioned exception of some Christian fundamentalist groups. A few years ago an interesting study on the effect of contact with aliens on religions was performed and put on the Internet with the title *The Extraterrestrial Sermons.*[2] The study develops scenarios in which an extraterrestrial probe answers a transmission broadcast from our planet in 1999 (an actual event). Three types of answer were considered: one simply showing that extraterrestrial intelligence exists, one telling us that they believe in some sort of god, and a third containing an atheistic statement. The effect of each answer was studied in the context of four different scenarios: one Muslim, two Christian, and one Hindu community. In each of the twelve resulting scenarios, the community's religious beliefs were strengthened or, in the last case, at least not undermined, by the message.

---

[2] http://www.richardb.us/project.html.

A Finite Universe and an Infinite God

The existence of extraterrestrial life, and even the vision of a universe rich with life, is more in agreement than in contrast with a truly religious vision of the world. One should remember that the main reason religious thinkers in the past could assert that the Universe is much bigger than that taught by Aristotelian philosophy—or even that it is infinite—was the consideration that to set limits to the extension of the Universe means setting limits to the power of God, and that an almighty God is manifested by the greatness of His creation. As we saw in the preceding chapter, many supporters of the plurality of worlds and of extraterrestrial life were mostly clerics, and Pope John XXI declared that not to believe in the possibility of a plurality of worlds is heresy.

Today there are still arguments on whether the Universe is infinite or not, but it has at any rate been ascertained that its size is so large in comparison with human scale as to be almost unimaginable. It can in fact be affirmed that even if it is finite, the Universe is to all intents and purposes infinite as far as humanity is concerned: humans will never reach more than a very small part of it (if any).

This is why many consider it unsatisfactory, from a religious point of view, that God should have created such an immense and complex Universe and then populated only one planet with life and intelligence. This objection, which obviously carries no weight for those who think that life is a meaningless accident in a universe with no purpose or finality, is important in a religious worldview that assumes that, first, life and then intelligence are the true goals of creation, or at least partial goals, since it is possible that intelligence is just one stage in the evolution toward some still unknown—and unimaginable—higher goal.

Undoubtedly a vision in which life is just an accident in a cosmos without purpose is much more removed from a religious view than the idea of a universe teeming with life. Epicurean physics, which expressly and coherently assessed the infinity of the Universe and the ample diffusion of life, was opposed by the Christian Church as a part of Epicurean philosophy—that is, for its atheism and materialism and not for its speculation regarding the infinity of the Universe or atomism (today we would say for its scientific leanings). We now realize that to support a scientific theory on the basis of religious arguments is as wrong as to pretend that the Holy Scriptures form a scientific text, but this is exactly what many supporters of the plurality of worlds did in the sixteenth and seventeenth centuries.

However, if a universe teeming with life is much more easily incorporated into a religious worldview than a universe devoid of life except for one tiny planet, religions are typically constituted by a number of structured beliefs and many find it difficult to incorporate into their theological structure new facts that their founders didn't consider. The stiffer the structure of a religion, the more difficulties it will have in accepting new ideas, such as the existence of extraterrestrial intelligences.

HINDUISM, BUDDHISM, AND OTHER ORIENTAL RELIGIONS

Buddhism[3] and Jainism are branches stemming from the trunk of Hinduism, and share many of its characteristics. They are very flexible in their theological structure, and therefore it is very easy for them to accept new ideas. One of the main points of these religions is their respect for life in all its forms, and surely the idea that life is widespread in the Universe is welcomed by their followers, many of whom will consider it to be obvious. Surely, to them, extraterrestrial living beings are as sacred as those on Earth and the reincarnation cycle should include them, too. As each one of us goes through numerous reincarnations, so do extra-terrestrial beings and the cycles probably interact with each other.

A central point of Buddhism is enlightenment. Tibetan Buddhism holds that only a small number of the innumerable forms of life existing in the Universe can attain enlightenment and, in a way, this is also obvious. You cannot really expect bacteria or even complex but nonconscious animals to be eligible. Edgar Martin del Campo[4] suggests that Mahayana Buddhism might imply a formulation similar to the Drake equation (see Chapter 4) to state how many species capable of enlightenment can exist in our galaxy.

The very flexible theological structure of Hinduism and Buddhism—so flexible that they can accommodate theologies in which myriad gods exist and others that do not accept the presence of any god at all—certainly facilitates things. Some Hindus go to the point of embracing UFO cults or putting the presence of extraterrestrials at the center of their faith.

---

[3] A 1998 statistic states that 13% of the world population is nonreligious; of the remaining 87%, 26% are Christians, 20% Muslims, 13% Hindus, 6% Buddhists, 6% follow Chinese folk religions, and 10% other minor religions such as Baha'is (0.1%), Confucianists (0.1%), ethnic religionists (4.2%), Jains (0.1%), Jews (0.2%), New-Religionists (1.7%), Sikhs (0.4%), Spiritists (0.2%), seventy minor world religions, and more than 10,000 national or local religions. (Source: *1999 Britannica Book of the Year*, p. 315.)

[4] Edgar Martin del Campo, "A Rare Opportunity," *Theology Journal*, vol. 41, no. 7, 1999.

JUDAISM

Judaism, like Christianity and Islam that derive from it, is a more structured religion, based on a collection of divine revelations (the Bible) and other books, such as the Talmud, collecting the teachings of the rabbis. Its followers cannot accept what is in contradiction with the Holy Scriptures, but things are not so clear-cut since their interpretation is not unequivocal on many points, and different schools of thought therefore exist.

A small minority that sticks to the letter of the scriptures hold that since the Bible does not explicitly mention other habitable planets or extraterrestrial life or intelligence, they cannot exist. This minority shares this belief with the above-mentioned Christian fundamentalist groups and think that their discovery would threaten religion and is likely to be a trick staged by the devil. Note that this minority does not deny the existence of spiritual beings, like angels and demons, who live on Earth but also in the heavens (though often it is not clear whether "heaven" represents a theological or a physical space).

Perhaps the problem with the "religions of the book" is their axiom that the entire Universe was created for the sake of humanity, and in the Talmud it is also stated that all angels and spiritual worlds exist for the same purpose. So it could be said that religions based on the Bible are too anthropocentric to accept extraterrestrial life and, above all, extraterrestrial intelligence.

Actually the difficulty is more general: in the ancient view of the world there was no difficulty in thinking that all that exists within the sphere of the stars has some usefulness to humankind, but what could possibly be the usefulness of a galaxy located billions of light-years away, not visible without sophisticated instruments, to a species living on this planet? In this way anthropocentrism becomes incompatible with the whole of modern science. A way out of this is the possibility that in the future humans will spread through the whole Universe, and there are some interpretations for which spaceflight is a necessary prelude to the Messianic age, but on that scale the idea of "usefulness to humankind" actually looks devoid of all common sense. The only reasonable solution is to interpret the term *human* in a wider sense, as "intelligent and sentient being." Anthropocentrism in this sense simply implies the existence of a very large number of intelligent species in the Universe.

In his letter to Galileo mentioned in the Preface, Kepler uses exactly this argument to demonstrate that Jupiter must be inhabited: its newly discovered satellites cannot be useful to human beings on Earth, so there must be other "humans," that is, intelligent beings, close by. Therefore Jupiter must be inhabited.

Another important point raised by the plurality of the worlds is "Original Sin." But since this is common to Judaism, Christianity, and Islam, it will be dealt with later.

ISLAM

Islam is based on the Bible but its actual spiritual guide is the Qur'an. In a way, Muslims tend to stick more to the Holy Scriptures than Jews and Christians, but the very size of the community and the absence of a centralized religious authority produce quite diversified interpretations, at least on topics that are not at the core of the religion. Many Islamic scientists interpret some Suras of the Qur'an in favor of the plurality of words. Actually both spiritual creatures (angels, jinns) and material ones (animals, humans, collectively mentioned as Dabbatun) are mentioned and it appears that both live on Earth and in the heavens [e.g., Sura 16, verse 49: "And to God doth obeisance all that is in the heavens and earth, whether moving (living) creatures or angels..."[5]], with the expression "heaven and earth" indicating the whole Universe. In the same interpretation it is stated that both jinns and humans can be damned or saved, depending on their actions, so that the final judgment will deal with all intelligent and sentient beings, or at least those endowed with free will—humans and aliens alike.

In these interpretations, the Qur'an also foretells a contact between humans and aliens (or better, in this context, between humans from planet Earth and humans from other planets), and as a consequence an extraterrestrial message will be proof that the Qur'an is inspired by God and surely not something giving discredit to Islam.

In the Web site mentioned in footnote 5, the case of Ibn-e-Abbas is related. He was one of the Companions of the Prophet and one of the great scholars of the Qur'an. He believed in the plurality of the worlds and that the inhabitants of other planets had a revelation from God, just as humans had done on Earth. He even said that they have a prophet like Muhammad. However, he did not mention his ideas to others very often,

[5] The translation of the Qur'an is an open problem. Muslims hold that the Qur'an must be read in Arabic and no translation can be relied upon. Here the translation is that supplied by Samir Khalid Munir on the Web-site exobiologist@aliens-in-quran.com.

as he feared that he may shake their faith by telling them things they found hard to believe.

Anyway, it must be said that Muslims are often quite suspicious of theological speculation and tend to put much more emphasis in their faith in divine revelation.

CHRISTIANITY

The problems facing Christians in dealing with the existence of extraterrestrial life and intelligence are similar to those we have already seen for Jews and Muslims, namely the anthropocentrism of the Bible and Original Sin, plus an even more serious problem, linked with the central event of Christian religion: redemption.

As already stated, a literal interpretation of the Bible may preclude the possibility of accepting the plurality of the worlds. A literal interpretation of the Bible, however, is in contradiction not only with this specific issue but also with almost all modern science. The trial of Galileo is very interesting from this viewpoint. At that time, the Catholic Church was supporting the Aristotelian view of the Universe, even if in itself it had little to do with the essence of religion, because it seemed in better agreement with the Holy Scriptures.

The conflict between the religious and scientific (or philosophical) views of the world was already old in Galileo's time. Many Christian Aristotelian philosophers (particularly Christian followers of Averroës such as Sigeri di Brabante), to solve the contradictions between the Holy Scriptures and the views of Aristotle, supported the "two truths" assumption asserting that philosophy and religion belonged to two different planes, truly independent and not subordinated to each other. When they lead to contradicting results, both must be assumed to be correct, each in its own domain.

To Galileo this conclusion seemed unsatisfactory. He thought that there must be only one truth, irrespective of the way in which we achieve it. He stated that nature and Holy Scriptures both derive from God, and their study should lead to the same conclusions. However, when they are apparently in contrast, we must remember that the Bible was expressed in a form that could be understood by the people to whom the revelation was directed, while in nature the will of God is expressed in its pure form. So, if anything, the results of scientific endeavors can be used to better understand and interpret the Holy Scriptures. He felt it his duty to express this idea to the highest authorities of the Church, to convince them that they should not entrench themselves in a scientifically incorrect position. When he was summoned to Rome, he went in this spirit and full of hope and pursued his "cultural policy," particularly after the election of Pope Urban VIII, an open-minded scholar who protected independent philosophers. The final outcome of his trial was even sadder for him, as he felt he belonged to the Church and feared that the victory of its most conservative exponents would cause severe harm not only to science, but ultimately to the Church itself. Galileo's trial was not so much a clash between science and religion as between two different views of religion.

At present, Christian theologians, or at least Catholic theologians, agree that the revelation in the Holy Scriptures was "suitable" for the people who received it, and must be interpreted taking into account the cultural, and therefore also the scientific and material, environment in which they lived.

Christian churches and religious authorities in general—with the notable exception of the fundamentalist groups present in every religion—have long since abandoned the idea that the Holy Scriptures can be used to defend scientific ideas and they now keep the scientific and religious spheres rigorously separated. It should also be remembered that the importance of fundamentalism is often overestimated, in the sense that the activism of their followers makes fundamentalist ideas appear much more widespread than they actually are. Such activism, however, also represents a danger that must not be underestimated, as it can give these groups a far heavier political and social weight than their actual numbers could ever gain them.

It is nevertheless clear that the discovery of life—above all, of intelligent life—outside our planet would have consequences for our vision of the world deeper than those due, for instance, to the discovery of new elementary particles or even to a radical change of a paradigm in some scientific discipline. When science deals with *life* it touches something many believers feel belongs more to the sphere of religion than to science. It is no coincidence that the most recent clash between science and religion was caused by the theory of evolution, only recently accepted by the Catholic Church and still opposed by some other denominations.

The Problem of Original Sin

Original Sin is described in the first part of the Bible, and is therefore one of the bases of the three great monotheistic religions: it essentially consisted of an act of deliberate rebellion of man against the will of God. This act resulted in the entry of evil into the world and in the mortality of humans and all other creatures.

The problem that would be created by the discovery of extraterrestrial intelligence is not altogether new. When Christopher Columbus brought back some natives from his trips to the new continent, it caused no great problems since there had been no doubt that the Indies, thought to be a part of Asia, were inhabited. But when it became clear that the continent discovered by Columbus was a separate land mass, the idea that its inhabitants constituted a human breed that had nothing to do with the descent of Adam and Eve began to be discussed.

There were some who doubted that these "new" people were even human. In Christian terms, this meant doubting that they had an immortal soul. Such doubts had few practical consequences, since the unity of humankind was instinctively clear, even if in the nineteenth century there were still theories that traced the origin of the various human races back to different evolutionary lines that were well outside the Christian world. Today the oneness of the human species and the marginality of the differences among the human races is a scientific certainty accepted by everybody, but there is no doubt that problems of the same type will re-emerge with the discovery of extraterrestrial intelligence.[6]

This problem was already well understood at the end of the seventeenth century. Bernard le Bovier de Fontenelle, in the preface of his book *Conversations on the Plurality of Worlds*, wrote:

> When I say that the Moon is inhabited, you immediately think of men like ourselves, and then, if you are a bit of a theologian, you are instantly full of problems. ... The men who live on the Moon are not sons of Adam. ... The inhabitants I put on the Moon were not like men in any respect.

Fontenelle thinks he has solved all the theological problems by stating that the extraterrestrials are not men, but things are not that simple. In the very instant in which we assume that there are other intelligent, conscious beings in the Universe, to say that they are human or not human is just a question of words, and the essence of the problem remains.

According to the letter of the Scriptures, Original Sin took place in history or, better, at the beginning of humanity's history. Now it seems reasonable that an act of deliberate disobedience presupposes the existence of an intelligent creature endowed with free will. It could in fact be said that Original Sin is a consequence of the evolution of an intelligent species on this planet. The existence of other intelligent species therefore initiates a serious theological problem: are extraterrestrial intelligences also

---

[6] Racist theories assessing the different origins of the human groups were the basis of political movements of the past, which, unfortunately, still have followers. Their danger cannot be underestimated.

involved in such acts of rebellion against God? And, should a multiplicity of intelligent species exist, is it possible that some of them have not been touched by Original Sin? If the answer is in the affirmative, it would mean that, in the Universe, there are species not touched by evil and death—creatures that could be defined as angelic, even if they belong to the same material world as the humans of Earth. If, instead, Original Sin is general and affects every intelligent species, is it still possible to define such behavior as an act of voluntary disobedience?

Actually, Original Sin causes problems that go well beyond compatibility with the existence of extraterrestrial intelligence and, as Father Pierre Teilhard de Chardin[7] explicitly noted, it gives problems of compatibility with the whole vision of the world that science has developed in the last century and a half, particularly with the scientific explanation of the origin and the evolution of the Universe and life on Earth (see Chapter 3).

The French Jesuit noted that as death has been a characteristic linked to life since its beginning, it is necessary to locate the Fall in the Precambrian period (today we know that life on Earth started some billion years earlier, and therefore it has to be located almost four billion years ago, shortly after the formation of the planet) and not in the time of the evolution of our intelligent species. But surely it is unthinkable that the first, primitive life-forms, simpler than bacteria, could have been able to commit a conscious act of rebellion against the divine will. The existence of other forms of life—many of which are much more ancient than terrestrial life, but surely mortal—does not change the substance of the problem, it just moves it back in time and transfers it to another place.

Teilhard's suggestion is to consider Original Sin as a "reality of trans-historic order," an event that cannot be located in a particular place and time, but affects all objects in the Universe "from the first to be formed to the most distant of the galaxies." He develops this hypothesis according to two possible schemes. The first follows a Christian cosmogenesis of a more traditional type, and can be represented by the scheme in Figure 2.1(a), taken from the work of J. Carles and A. Dupleix.[8]

God (G) instantly creates a perfect human creature (A_1, the first Adam): this is the Eden phase. The cone of involution (I) represents the Fall (of Original Sin). A precosmic phase of involution, which produces the multiple (M), is thus present. A cosmic phase of evolution (EV) in

[7] Pierre Teilhard de Chardin, *Meditations on Original Sin*, Paris, November 15, 1947, reported in J. Carles and A. Dupleix, *Tehilard de Chardin, Mistico e Scienziato*, Edizioni Paoline, Milano, 1998, pp. 268–278.
[8] J. Carles and A. Dupleix, op. cit., p. 273.

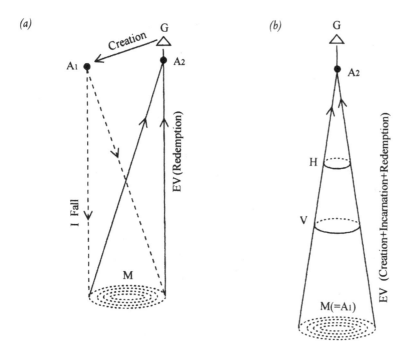

FIGURE 2.1 *Possible schemes proposed by Pierre Teilhard de Chardin to insert Original Sin within a cosmogenesis compatible with what has been ascertained by modern science.*

which our history is situated then follows, and that tends to the second Adam, that is to Christ (A_2).

Teilhard, however, found this solution unacceptable for a number of reasons: for instance, the extracosmic phase is not necessary and the Fall seems to be an unnecessary and unlikely event. He therefore proposed the solution sketched in Figure 2.1(b). The multiple (M) is the original form of the Universe since the beginning, and the creative act of God consists of a gradual process of organization and unification. Original Sin is no longer an isolated act: it becomes a state, a generalized presence of disorder that implicates, for life, the existence of pain and death, and for the human condition, the existence of sin.

In the cone of the evolution of the Universe (EV) there is a first prebiotic phase (chemical evolution) up to level V, then a phase of biological evolution leading to the appearance of humans (level H). From such a level the history of humans begins, with the entry of liberty in the cosmic order, to the arrival of Christ, the second Adam (A_2), and then to the final "recapitulation" of the Universe in God (G).

Clearly, both these attempts to interpret Original Sin within a universe similar to that described by modern science are perfectly compatible with the existence of both a plurality of inhabited worlds and a plurality of intelligent species.

The Problem of Redemption

If the problem of Original Sin is common to the monotheistic religions based on the Bible, that of redemption is typical of the Christian religion only.

The event of the incarnation of Christ and the consequent redemption happened at a well-determined time and in a well-determined place; hence it must be situated in the history of humanity. It does not conflict with the scientific view of the world, at least if we limit ourselves to considering our planet. The situation may be quite different in the case of a multiplicity of intelligent species inhabiting the Universe. The main question that extraterrestrial intelligence poses to the Christian theology is whether the salvation brought by Christ with his incarnation on this planet only concerns the human beings of the Earth or all the possible intelligent beings living in the cosmos. The question is old and dates back at least to the commentaries to Aristotle of the fifteenth century, and the standard answer was that it concerns the whole Universe, but at that time it was just an academic question, since Aristotelians did not believe in the plurality of worlds.

This solution gives rise to perplexities today. It would give Earth a central role, a geocentric view that started waning with Copernicus and eventually disappeared completely in the modern conception of Earth as a "standard" planet located in a "standard" galaxy of this Universe. Of course the centrality that Earth would assume if it should turn out to be the only planet inhabited by an intelligent species is of a totally different type: in this case Earth would be something special, but for this reason we, as intelligent observers of the Universe, could not by definition be in any other place.

Moreover, for reasons that will be shown later, it is extremely likely that if other intelligent beings exist, the majority of them are much more ancient than we are. To suppose that ours is the oldest species would mean giving back to humanity a position of privileged observer that, as we have just seen, appears to be incompatible with the current worldview.

In such a case it would also be difficult to understand why salvation has not been brought into the world much earlier, especially since many species may well have been extinct billion of years before the incarnation of Christ on Earth.

Finally, if it should actually occur in the future that we come into contact with other intelligent beings, this kind of perspective would imply a sort of missionary effort performed by the humans of the Earth—an idea that causes perplexity at the very least.

The alternative is that a multiplicity of incarnations and redemptions have occurred (and will occur in the future), and that the Son of God incarnates from time to time in beings of the various intelligent species when they reach a determined stage in their evolutionary history or, better, in the history of their salvation. Actually, such a hypothesis is perfectly compatible with divine omnipotence and with the infinity of God's love for humanity (intending with such a term any community of intelligent beings).

The hymn writer, Sydney Carter, wrote

> *Who can tell what other cradle*
> *High above the Milky Way,*
> *Still may rock the King of Heaven*
> *On another Christmas Day?*[9]

This perspective was ridiculed in 1794 by Thomas Paine in *The Age of Reason*: "The Son of God, and sometimes God himself, would have nothing else to do than to travel from world to world, in an endless succession of death, with scarcely a momentary interval of life."

However, while this is little more than a joke, since it has no theological relevance, the idea of multiple redemptions gives rise to perplexities. To many it seems that this automatism (one intelligent species—one incarnation) is too schematic and doesn't take into account the infinite freedom of an omnipotent God. When Sir John Polkinghorne, the physicist and theologian, was asked about this problem, he answered that God "will do what is necessary" in the different situations.[10] On the same line, Father Georges Coyne, the director of the Specola Vaticana (the Vatican Observatory), observed that "God chose a very specific way to redeem human beings," suggesting that in other circumstances He could do otherwise. In any case, Christian theology suggests that humans living a virtuous life can be saved even if they do not know the word of God. As is the case for Muslims, all good intelligent and conscious humans will also be saved, their planet of origin, species, or exterior aspect being immaterial.

---

[9] S. Carter, *Every Star Shall Sing a Carol*, Copyright 1961, Stainer & Bell Ltd., from the Web site http://www.ufoevidence.org/documents/doc1711.htm.
[10] Quoted in *The Observer*, August 11, 1996.

BIOCOSMIC THEOLOGY

The problems seen above do not actually involve any religious dogma; they simply require new theological elaboration—something that has always happened in the history of all religions. Many theologians and scientists near to the churches have expressed opinions supporting the existence of extraterrestrial life and intelligence; for example, in the words of Father Angelo Secchi, director of the Specola Vaticana in the second half the nineteenth century (a contemporary of Schiaparelli): "Life fills the Universe, and intelligence must be associated with life; and like beings inferior to us are innumerable, so in other conditions beings immensely more advanced than ourselves can exist."[11]

In this statement Father Secchi puts forward the idea, which is in accord with what we now think, that many of the intelligent species existing in the Universe are more advanced than ourselves. Human beings no longer play the role of the most advanced (and therefore closest to God) beings of creation, second only to angelic creatures. They have lost a role that had long been theirs in the Christian view of the world.

Another supporter of the existence of extraterrestrial intelligent life was St. Maximilian Kolbe. Not only was he convinced of its existence, he thought that a sort of link exists between all rational beings and that, in the future, humanity will go in person where only its eyes can now reach, and meet intelligent extraterrestrial beings. In the meantime it is necessary to improve the instruments of observation. In 1915 he suggested using a space vehicle capable of traveling outside the atmosphere to perform detailed astronomic observations of the planets to search for extraterrestrial life. No doubt these statements recall, in a Catholic key, the basic beliefs of Cosmism, even if there is no evidence that he knew about that movement.

More recently, the Dominican Reginaldo Francisco reports a conversation between the philosopher Jean Guitton and Pope Paul VI, in which the former affirmed

> our descendants will perhaps come into contact with other "reasoning" species. What will happen then? Here is the way I formulate the problem. First hypothesis: such reasoning beings won't have known Jesus Christ. Second hypothesis: the Verb will have been "proportionate" to them, i.e. it will have been expressed to them under other forms. In this case it will be impossible to

---

[11] Quoted in G. Schiaparelli, "Il pianeta Marte," reprint of issues 5 and 6, 1893, of the journal *Natura ed Arte*, reported in P. Tucci et al. (eds.), *Giovanni Virgilio Schiaparelli, La Vita sul Pianeta Marte*, Mimesis, Milano, 1998.

speak of a Church. The terminology of this Cosmic Church should find something equivalent in our language. The truths of faith—as happens for the scientific and philosophical truths or when we discover an unknown language—will then be transferred. Some prayers, such as the *Common Preface*, in which the orator is said to be in relationship with the Celestial Choirs, and some obscure books, such as the *Apocalypse*, lead us to think that the Catholic Church is as vast as the worlds it possesses. But the Catholic Church is the Church of all the worlds. We must therefore extend the beautiful word "Catholic" to all the Universes. The revelation of Christ embraces all humanities.[12]

Francisco informs us that such a hypothesis, though it was not officially approved, seemed reasonable to the Pope. It should be noted that Guitton, essentially remaining within one of the above-mentioned hypotheses, goes as far as to say that the revelations to other humanities may be expressed in a substantially different way, to the point that it would no longer be possible to speak of a Church, but that in spite of this a sort of communion with a Cosmic Church exists among all intelligences of the Universe. The term *Catholic* is extended to include not only all human beings on Earth but all intelligent beings, to whom the term *human* should be extended, becoming synonymous with intelligent creature.

Reginaldo Francisco proposes the foundation of a "biocosmic" or "space theology," defined as "the science that, inspired by Revelation and using the experimental and scientific data, studies and develops a new view of material and spiritual being in the cosmos, in relationship to God and his people."[13]

Biocosmic theology is therefore based on the assumption that life and intelligence are present more or less throughout the Universe, even if it would actually be enough for a single extraterrestrial intelligent species to exist in the whole Universe. It is also probably necessary to assume that contact—at least in the form of an exchange of information, but above all a direct encounter between intelligent species—is possible.

Without contact, in fact, not even the certainty of their existence could be reached and little would change with respect to the present situation. And it is clear by now that contact will be possible only if the number of intelligent species in the Universe is enormous.

Today we think that the number of galaxies is on the order of hundreds of billions and therefore, assuming that intelligent species are

---

[12] Reported in Reginaldo Francisco, "Possibilità di una redenzione cosmica," in F. Bertola et al. (eds.), *Origini, l'Universo, la Vita, l'intelligenza*, Il Poligrafo, Padova, 1994.

[13] Reginaldo Francisco, "Prolegomenos para una teologia bio-cosmica," in *Libro Anual*, Lima, Facultad Teologica y Civil, 1968, pp. 81–103.

distributed more or less evenly in space, their number has to be of the same order, so that at least one exists in each galaxy, or, better still, so that every galaxy contains at least one intelligent species throughout the *duration* of the Universe. Nothing can be said of their distribution in time, since nothing is known either about the duration of the Universe (various hypotheses exist, but no certainty has been reached) or about the life span of intelligent species, whose duration could also be vastly different between one and any other. It is reasonable to think that for contact with a unilateral exchange of information to be possible, the number of intelligent species in the whole Universe (meaning in the whole space *and time* of the Universe) has to be in the thousands or even millions of billions, which may be reasonable given its size (both in time and space).

Teilhard de Chardin, for instance, thought it possible that other biospheres could exist, in addition to the one on Earth, and that they might give origin to a *noosphere*, a term with which he indicated all matter constituting intelligent beings or, as he put it, *hominized* matter. Yet, as he considered it impossible, or at least highly unlikely, that there would ever be contact between them, he didn't deal with the problem, concentrating all his scientific and philosophical studies on the biosphere and the noosphere of the Earth.

> The Stars are very likely scattered in space without possibility of communicating for the very reason of containing each one a special soul, the soul of the people that multiply on their surface, the collective soul of all those that the cosmic isolation compresses in love and effort, up to the birth of a mysterious organism, originated from their consciousness.[14]

Finally, it should be noted that many people tried to interpret some passages in the Holy Scriptures in an extraterrestrial light (or perhaps it would be better to say in a UFO light). Actually it is even too simple for the fertile imagination of the UFOlogists and such authors as Eric Von Daniken to attribute, for instance, the destruction of Sodom and Gomorrah to extraterrestrial spaceships and weapons, or to think that Elijah was the more or less willing passenger of a flying saucer. Even Jesus was considered an alien and His ascension was reduced to a sort of "Beam me up, Scotty." Since not only the Holy Scriptures but almost all ancient texts (including many sculptures, sketches, and other forms of recording information) are subjected to the same delusional treatment, and since the

---

[14] From a writing of Pierre Teilhard de Chardin, January 15, 1918, reported in J. Carles and A. Dupleix, *Tehilard de Chardin, Mistico e Scienziato*, Edizioni Paoline, Milan, 1998, p. 224.

question does not really have any theological or religious implications, it will be dealt with in Chapter 5.

However, it is perhaps worthwhile quoting, with all due caution and doubt, verse 10.16 of the Gospel of St. John: "I also have 'other sheep' that are not of this fold; also these I have to lead, they will listen to my voice and they will become a single flock and a single shepherd."[15]

Actually no one has ever suggested that the Evangelist was speaking of extraterrestrial "sheep," but the verse helps one to presume that if extraterrestrial, intelligent, and conscious beings exist, the religious vision is to consider them as equal to humans, and Christians should consider that they are endowed with an immortal soul and are therefore "children of God and Brothers in Christ." The community of intelligent beings is therefore a single flock, whose fold is the whole Universe. In case an encounter with other intelligent beings is possible, they must be considered human beings to all effects, with the moral obligations that this implies.

However, if Original Sin applies to the whole Universe—and evil therefore pervades the whole cosmos—the fact that extraterrestrial life-forms are intelligent and endowed with free will also makes them human in the ability to do evil. Not only evil but error too—intending this term in the widest possible sense—is therefore connatural with the human condition. If, in this context, contact with other intelligent beings belonging to a species more advanced than ourselves should allow us to learn much in the scientific and technological fields, great caution will be required to ensure that we do not indiscriminately accept views of life and the world that are alien to our nature.

We must surely be disposed to learn from everybody, but we should not assume that technologically more advanced civilizations are closer to the knowledge of the ultimate truths. Here on Earth, the decadence of civilizations that came into contact with technologically more advanced communities was nearly always accompanied by the nondiscriminating acceptance of models brought by the latter. Even outside of religious views, one of the basic ideas of Cosmism was that every intelligent species must bring its specific sensibility and worldview as an original contribution to global civilization: a flattening on models brought from outside would only produce a general cultural impoverishment.

---

[15] The emphasis of "other sheep" is by Reginaldo Francisco, who quotes the verse in "Possibilità di una Redenzione Cosmica," in F. Bertola et al. (eds.), Origini, l'Universo, la Vita, l'Intelligenza, Il Poligrafo, Padova, 1994. Actually it is most likely that the Evangelist was referring to pagan peoples.

The above considerations can no doubt be accused of anthropomorphism: it is quite possible that when we discover extraterrestrial intelligent life-forms, they will be so different from us that these ideas will seem completely inadequate—we may even have difficulty recognizing them as such.

Nevertheless, a certain degree of anthropomorphism is perhaps reasonable from a purely religious viewpoint—and even more so from the point of view of monotheistic religions derived from the Bible. After all, the Bible clearly states that man was created in the image and likeness of God and there is no reason to think that the phrase has no general application. While in a purely scientific view it is possible to think of intelligent forms of life extremely different from each other, believers tend to see God as a unifying factor, at least as far as intelligence is concerned. The morphology, the biochemistry, and the evolution of beings may be arbitrarily different, but common lines, and therefore a common ground for understanding, should exist where the logical and spiritual spheres are concerned.

Everything that has been said in this chapter about a religious view of the Universe is, of course, not a necessity but just a possibility, partly because theology started to formulate problems of this type too recently to be able to perform a satisfactory elaboration. Believers, therefore, must also be ready for a future of surprises and for realities that, as always, will go beyond the wildest human imagination.

3

THE
ASTROBIOLOGICAL
PERSPECTIVE

A NEW SCIENCE: ASTROBIOLOGY

EXTRATERRESTRIAL life, its beginnings, and its evolution are the subject of astrobiology.[1] Some scientists, critical of this new science, expressed doubts about its scientific nature and defined it as

---

[1] Strictly speaking, the branch of astronomy studying the conditions for the development of life and trying to define general laws for the evolution of life in the Universe should be named bioastronomy. Astrobiology (sometimes referred to as exobiology) is the branch of biology studying extraterrestrial forms of life or at least performing biological experiments on other celestial bodies (e.g., the experiments performed on Mars by *Viking* landers). Recently, particularly in the United States, the term *astrobiology* started to be used in an extensive way. Following this practice, in this book only the term *astrobiology* will be used.

the only science that does not have a subject, since there is no evidence that extraterrestrial life actually exists. This is in fact a groundless objection, since astrobiology studies the conditions needed for the development of life and tries to define general laws for evolution from the simplest to the most complex forms of life on the basis of the data that astronomy and astrophysics are slowly accumulating regarding the conditions prevailing on the various celestial bodies.

In the last 40 years humanity has taken its first, still uncertain steps off its own planet, with human beings setting foot on the nearest celestial body and automatic probes exploring many other planets and satellites of the Solar System. Space probes have flown close to all the major planets (except Pluto) orbiting the Sun, and to some of their satellites, and objects built by humans have even landed on some of them.

Space technology allows us to deploy telescopes and other astronomical instruments outside the Earth's atmosphere, yielding much more detailed images than those taken from the surface of the Earth. They also make it possible to study wavelengths that cannot cross the ocean of air surrounding us. Besides optical astronomy and radioastronomy, the fields of infrared, x-ray and gamma-ray astronomy now look very promising, and the result of this research has been a revolution in our knowledge of the Solar System and other celestial bodies. Today we are aware of a very different Universe from the one we observed just 15 or 20 years ago.

This revolution forced scientists to drastically cut back hopes of finding extraterrestrial life, at least in our Solar System. The possibility of one day colonizing other planets also seemed to fade away. The nearest planets were found to be much more hostile to life than was previously thought, and this spread the idea of a universe, or at least a solar system, completely devoid of any form of life. Biologists are generally much more pessimistic than astronomers on this matter, perhaps because knowing in detail the complexity and fragility of life, they realize better than others the difficulties that can prevent life from developing.

Astrobiology received a great impetus in the last decade of the twentieth century, and with a progressively deeper involvement of the space agencies, and above all of the National Aeronautics and Space Administration (NASA), the picture began to change once again. Very resilient forms of life were found on Earth, and less hostile environments were discovered in the Solar System. Planets orbiting other stars (*exoplanets*) were discovered too. The idea that life is very common in the Universe once again attracted many scientists; the whole idea of a *biological universe* gained momentum.

TIMES OF THE UNIVERSE

The large majority of scientists think that the Universe began about 10 to 15 billion years ago with a huge explosion, the so-called *Big Bang*.[2] The idea that the Universe began in a single event was first proposed by Abbot Georges Lemaitre, an astronomer at the University of Liège, in the 1930s. Lemaitre combined the results obtained by Alexander Friedman—who worked on the equations of relativistic cosmology—with Edwin Hubble's discovery of the expansion of the Universe, and formulated the hypothesis that the origin of the present Universe can be traced back to the explosion of a "primordial atom" (according to his definition) containing all matter and energy. The idea was reformulated at the end of the 1940s by Russian-born American physicist George Gamov who, working from Lemaitre's hypothesis, inferred the existence of background radiation, characterized by a temperature of a few degrees above absolute zero, that should uniformly permeate the Universe.

In 1965 Arno Penzias and Robert Wilson, two engineers at Bell Laboratories, detected background radiation at 3 degrees above absolute zero (3 K) and this was considered, and it still is today, one of the most convincing proofs of the *Big Bang* theory. The catchy name had been introduced as a joke by theoretical physicist Fred Hoyle in a BBC radio program in 1948, and had such success that it is still used today. Hoyle, until his death in 2001, was one of the few scientists who opposed this theory, which over the years has come to be regarded more and more as conclusive. The *Big Bang* can be said to be one of the basic paradigms of today's physics.

Alternative theories—that is, theories that try to explain the shift toward the red end of the spectrum of light from the most distant galaxies, discovered by Hubble, without resorting to an expanding universe—do exist, but currently they have little following. There is no doubt, however, that some points need to be further clarified and that the basic hypothesis of Lemaitre and Gamov has undergone, and is still undergoing, changes and improvements.

New experimental results require new theoretical elaborations (such as the recent observation that the expansion is accelerating, instead of slowing down as the initial theory had predicted), but the majority of cosmologists are convinced that the basic points have now been established and that the theory is substantially correct.

The initial stages of the *Big Bang* took place at a fantastic speed and can

---

[2] A figure often considered as most likely is 13.7 ± 0.2 billion.

now be reconstructed in some detail. It is possible to go back to a very small fraction of time (much, much less than a second) after the initial instant. It is, however, impossible to go back beyond a certain point, since in the earliest phases conditions were such that the laws of physics were different from those that currently govern the Universe. Besides, it would have little meaning to wonder what the Universe was *before* the *Big Bang*: if it really is a "singularity," it would have canceled any traces of a possible previous history.

Theoretical elaboration is proceeding apace. Andrei Linde, for instance, suggested in the 1980s that our Universe is enormously greater than is generally accepted, and is just one of many universes that originated from "quantum vacuum fluctuations." Apparently, no contact is possible between these universes, and therefore the theory seems to allow for no experimental verification, at least in no direct sense.

One must be careful not to identify the *Big Bang* with Creation and not to draw evidence or clues from cosmology for use in the theological field. One must not confuse the scientific and the religious planes as Pope Pius XII did, for example, when he interpreted the *Big Bang* as God's creation of the Universe. But it is equally questionable to affirm, as Stephen Hawking has done, that God is a nonnecessary initial condition for explaining the Universe.

Without going into too much detail (details that are still keeping cosmologists busy), it can be stated that at the very beginning the temperature and density were so high that everything was in the form of energy. Then, with the initial expansion, energy condensed into elementary particles, protons, and neutrons, and electrons began to form. For a short time, about half an hour after the initial instant, the temperature was low enough to allow some protons to react with each other to form helium nuclei.

That phase of *cosmic evolution* brought the formation of various chemical elements that are still present today. Yet, for about 300,000 years the Universe was just an expanding and cooling mixture of energy in the form of radiation, elementary particles, and hydrogen and helium nuclei. With further expansion and cooling, the conditions came into being for the protons and helium nuclei to capture electrons and form atoms. Still, in the rapid expansion immediately following the explosion, the Universe consisted mainly of hydrogen, with a certain quantity of helium.

The expansion continued, and slowly zones with a slightly higher density began to form; they then gravitationally attracted other matter, giving rise to the first galaxies. Inside the galaxies clouds of very thin gas continued to split into smaller clouds, which tended to collapse toward their center, slowly giving birth to the first stars. Within one hundred million years, the first galaxies and the first stars had formed. Continuing to contract gravitation-ally, and therefore warming up, they reached temperatures high enough to start the first nuclear reactions. The first part of the process of chemical evolution, based on a chain of nuclear reactions that resulted in the formation of all elements heavier than helium, had begun: from lithium to uranium, passing through oxygen, carbon, silicon, and all the others.

Stars have a life cycle through which they progress at different speeds, depending on their mass: those similar to the Sun have a life span of numerous billion years, while smaller ones may live for a longer period. In the Universe there are stars of all sizes, starting from about ten times the planet Jupiter up to giant stars whose mass is thousands of times that of the Sun. The largest stars burn their nuclear fuel quickly and, some hundred million years after their birth, explode very violently, scattering their material at distances of many light-years.

Other material, containing the products of the nuclear reactions taking place in the stars, is continuously expelled into interstellar space in the form of stellar wind. In this way, some hundreds of million of years after the *Big Bang*, atoms of elements that later would be essential for life began to diffuse into space. Slowly, the material expelled by the stars mixed with other interstellar material, giving rise to nebulas and then contracting into protoplanetary disks, to give life to new stars. But contrary to the formation of earlier generations of stars, this time the presence of heavy elements in the nebulas allowed the formation of rocky planets.

These newer stars also began to synthesize heavy elements, scatter them into space thanks to the stellar winds, and eventually explode at the end of their life cycle. This process was repeated many times in the billions of years that passed from the *Big Bang* to the formation of the protoplanetary nebula that gave origin to our Sun and Solar System. The nuclear synthesis of the heavier elements undoubtedly required very long time periods; there were several generations of stars before the nebulas from which new stars were born were sufficiently rich in heavy elements to initiate the formation of rocky planets, potentially endowed with the elements required for life, or at least life as we know it.

In addition to hydrogen—an essential constituent of water and an element present in the Universe since its first stages—there were other elements that we commonly associate with life, such as carbon, oxygen, and nitrogen, and many light elements that started to form at an early stage in the history of the Universe. The forms of life that developed on Earth also used many relatively heavy metals (for example, iron, an essential component of hemoglobin), which, even if only present as traces in living organisms, were essential.

Among the so-called biogenic elements—that is, elements necessary for life—carbon plays a special role. The structure of its electronic shells permits the formation of long chains and complex ring structures, giving rise to the complexity of what we call organic chemistry and to the possibility of an almost endless variety of compounds essential for life, so much so that the form of life that evolved on Earth is commonly referred to as "carbon based."

Our galaxy began to form around 10 billion years ago and, within it, after a few generations of stars had gone through their life cycles, our Solar System began to form around 5 billion years ago. At that time enough heavy elements were available for the formation of rocky planets that were able to host life and produce living matter itself. We do not actually know whether the 5–10 billion years that passed from the beginning of the Universe to the formation of the Solar System were indeed necessary for life to start (and all systems older than that are therefore sterile), or if planets able to support life could have formed billions of years earlier.

Although it is likely that the first generation of stars was unsuitable for life, the elements needed for life were probably present in planetary

TABLE 3.1 *Relative abundance of the most common elements in the Universe in general, in the living beings, and in the Earth's crust. (From C. Cosmovici, "Comete e bioastronomia," in F. Bertola et al. (eds.), Origini, l'Universo, la Vita, l'Intelligenza, Il Poligrafo, Padova, 1994.)*

Element	Universe (%)	Living matter (%)	Earth's crust (%)
Hydrogen	87	16	3
Helium	12	0	0
Carbon	0.03	21	0.1
Nitrogen	0.008	3	0.0001
Oxygen	0.06	59	49
Neon	0.02	0	0
Sodium	0.0001	0.01	0.7
Magnesium	0.0003	0.04	8
Aluminum	0.0002	0.001	2
Silicon	0.003	0.1	14
Sulfur	0.002	0.02	0.7
Phosphorus	0.00003	0.03	0.07
Potassium	0.000007	0.1	0.1
Calcium	0.0004	0	0
Argon	0.0001	0.1	2
Iron	0.0007	0.0	18

systems much older than ours. Yet it is interesting to note that the Sun appears to be exceptionally rich in heavy elements. If this is really an anomaly, rocky planets and life could be rarer than is usually thought. The relative abundance of the most common elements in the Universe, both in living beings and in the Earth's crust, is shown in Table 3.1. The first column is mainly based on the composition of the Sun, which is taken as typical of stars in general.

THE ANTHROPIC PRINCIPLE

The term *anthropic principle* was first proposed in 1973 by Brandon Carter during the symposium *Confrontation of Cosmological Theories with Observational Data*, held in Kraków to celebrate Copernicus's 500th birthday, as if to proclaim that humanity does hold a special place in the Universe after all. In his contribution, "Large Number Coincidences and the Anthropic Principle in Cosmology," Carter remarked: "Although our situation is not necessarily *central*, it is inevitably privileged to some extent."

The anthropic principle in its most basic form states that any valid theory of the Universe must be consistent with our existence as carbon-based human beings at this particular time and place in the Universe. In other words, "If something must be true for us, as human beings, to exist, then it is true simply *because* we exist." Attempts to develop cosmological theories from this principle have led to some confusion and much controversy, so we will start from the beginning.

The Universe, as described by *Big Bang* cosmology, seems to be very old, if we compare its age with human life spans or even with historical times. But the age of the Universe is not much greater, in terms of orders of magnitude, than the geological time scales to which we are accustomed on our planet. The age of the Earth is about 4–4.5 billion years, our galaxy is only twice that age, and the Universe three times.

Those who have speculated on the future of the Universe have usually ended up deducing that it will exist much longer in the future than it has done in the past, and this applies to both of the likeliest scenarios: (1) the present expansion will be followed by a contraction and a final implosion (the *Big Crunch*, a hypothesis that now appears to be incompatible with the latest discoveries on the acceleration of the expansion of the Universe); and (2) the expansion will continue indefinitely toward the final thermal death, the terminal state in which entropy is maximum and the Universe consists once again of undifferentiated matter.[3] A future in the tens of thousands of billions of years is usually predicted: the Universe would therefore now be extremely young, in fact in its early evolutionary phase.

In any case, the Universe seems to have been carefully planned to evolve in the manner that has actually occurred—toward greater complexity. And the most complex objects the cosmos contains (or which we know about, since it is possible that even more complex systems exist somewhere in the Universe) are living beings and, even more so, intelligent beings. What such a statement actually means is that the physical laws and the values of the constants that appear in them seem to have been designed for producing systems of increasing complexity. Many examples are brought in to support this thesis. Here are just a few of them:

- *The initial speed of expansion of the Universe.* Had it been smaller, even by a tiny amount (one hundred thousandth of 1 percent), the Universe would have collapsed back on itself almost immediately, while if it had been greater, also by the same tiny amount, the expansion would have been so rapid as not to allow the formation of galaxies.
- *The values of the constants in nuclear physics (for instance, the relationship between the mass and the charge of an electron).* If the reaction that caused the formation of the nuclei of carbon C^{12} starting from three nuclei of helium He^3 were not resonant, the quantity of carbon produced in the supernovas would have been very small, in fact too small for the development of living beings. In the same way, the reactions that brought about the formation of helium immediately after the *Big Bang* could easily have converted all the hydrogen into helium, thus preventing the formation of long-lasting stars and the presence of water in the ensuing phases of cosmic evolution.[4] All these reactions are very sensitive to the constants mentioned above, and even very small differences from the values they indeed have would have prevented the Universe from evolving toward a great complexity.
- *The properties of water.* Water has a unique property, due to the characteristics of the interactions between the atoms that constitute its molecule: its density in the liquid state is higher than that in the solid state (ice floats on water). Without this anomalous property, it is very likely that life as we know it would not be possible.

---

[3] Today some scientists cast doubts on the hypothesis of the thermal death of the Universe, given for certain in the past, also in the case of a Universe expanding forever.

[4] The reactions are $He^4 + He^4 = Be^8$ and $Be^8 + He^4 = C^{12}$. (J.D. Barrow and F.J. Tipler, *The Anthropic Cosmological Principle*, 1986.)

It should be noted immediately that what has been said above is rather hypothetical, since no theory exists to explain the basic laws of nature and therefore we cannot know if the values of the constants appearing in these laws might have been different from those we observe or if they are determined by other laws. The values of the constants of physics might be ascribed to sheer chance, but the probability that the Universe would actually evolve to the complexity we now see is so low that this explanation is not reasonable.

But what does it mean if we say that the Universe is finely tuned for producing increasingly complex systems and for allowing the existence of life? The simplest interpretation is to reverse the statement, saying that we exist because the laws governing the Universe allow it; but this statement is trivial. No wonder critics of the anthropic principle call it a truism (or, more wickedly, a tautological statement).

A second explanation is that the physical laws have been planned by a Creator with a well-determined finality. This explanation is outside the field of science: science always tried to explain the observations of the physical world without resorting to external entities. Clearly it is possible that there are limits to this program and that, at a certain point, it is necessary to resort to entities external to the physical world, but science has the precise duty of pushing such limits as far as possible.

An explanation that does not resort to entities external to the physical world is the hypothesis that the Universe is much wider than what we see and that, in the expansion, zones have been created in which the physical laws are homogeneous but different from those of the nearby zones. Our visible Universe is a zone in which the characteristics are such as to allow our existence. Notice that this hypothesis does not violate the principle of mediocrity, assigning humanity a role of privileged observer. We are not in a very peculiar zone of the Universe by chance, we are in the only zone in which we can exist.

Instead of assuming that the Universe has many diversified zones, it is possible to forward the hypothesis of the existence of many universes or of a multiple universe (a multiverse). The result is similar, without the difficulty of imagining interfaces between zones with different laws. This idea has been recently developed in many directions, to the extent of imagining universes that are born within black holes or even universes purposely created by intelligent beings, living in a previous Universe.

Discussion of the anthropic principle is still going on, but many physicists think it has no grounds since it is the very lack of understanding of the basic laws of physics that makes us think they are particularly tuned for allowing our existence.

Chapter 3

CHEMICAL EVOLUTION

The nuclear reactions that took place inside stars in the past produced all the chemical elements we know, many of them in various "versions," that is in the form of isotopes with a different number of neutrons in their nuclei (it is the number of protons in the nucleus that characterizes an element). The clouds of gas in interstellar space contained a wide variety of elements. Atoms started to join to form molecules, from the simplest, the molecule of hydrogen (H_2) made by two atoms of hydrogen, to much more complex ones.

Spectroscopic and radioastronomic observations allowed us to verify that substances such as water (H_2O), carbon monoxide (CO), ammonia (NH_3), and many organic compounds—from methane (CH_4) and hydrocyanic acid (HCN) to more complex molecules—are abundant in interstellar clouds.

Small grains of silicates, of a size about one tenth of a micrometer (one thousandth of a millimeter) were formed. These grains, wandering in space for hundreds of millions of years, passing from the intense cold of the interstellar space to higher temperatures in zones close to the stars, were covered by molecules of various kind that produced chemical reactions leading to the synthesis of molecules like methilamine (CH_3NH_2), methylic alcohol (CH_3OH), and formaldehyde (H_2CO). The granules were covered not only by water ice, but also by other volatile substances, and by a layer of organic material.

This mechanism produced large quantities of organic material; an interstellar cloud of some light-years in diameter, for instance, could yield a quantity of organic material equal to the mass of the Sun.[5]

The process of chemical evolution in interstellar space continued until much more complex organic compounds were produced: polycyclic aromatic hydrocarbons and chains containing atoms of oxygen, silicon, and sulfur. Molecules containing atoms of phosphorus, an element that later would become an essential component of DNA, have also been discovered.

About 10 percent of the carbon in interstellar clouds is thought to be present in the form of complex molecules as polycyclic aromatic hydrocarbons.

Chemical evolution continued when interstellar clouds started to condense, forming protoplanetary nebulas and then celestial bodies such as planets, asteroids, and comets, which were formed out of them, at least in the only case we know in detail.

[5] Jean Heidmann, *Extraterrestrial Intelligence*, Cambridge University Press, Cambridge, 1995.

THE FORMATION OF THE SOLAR SYSTEM

About five billion years ago the protoplanetary nebula—which, as it contracted, would give origin to the Sun—started to form. In its gravitational collapse the nebula, initially a three-dimensional cloud of gas of (roughly) spherical form, gradually changed into a disk rotating around its axis. This transformation was due to the fact that the cloud had a non-negligible angular momentum. A well-known law of physics states that if no external force is applied, angular momentum is maintained; a planetary system thus has the same angular momentum as the nebula from which it originated, but during contraction its rotation accelerates because its moment of inertia decreases. The phenomenon is the same as that used by dancers and skaters to increase their speed of rotation by drawing their arms close to their body.

At the center a bulge formed, mainly of hydrogen and helium, while the disk had a varying composition: atoms formed in previously existing stars, grains of dust, and molecules of various kind. The Sun formed in the central part of the cloud: when the temperature reached values high enough to initiate nuclear fusion reactions, the new star began to glow. As the star began to get warmer and nuclear reactions started, smaller masses in the disk also began to condense; gradually, by subsequent aggregations, they would give way to planetesimals, small celestial bodies that later, by colliding with each other, would give origin to the planets. The lightest molecules escaped from the planets that were forming in the inner parts of the cloud and were much warmer than those forming in the external part. What remained were the rocky nuclei that gave origin to the four terrestrial-type planets (Mercury, Venus, Earth, and Mars).

The planets that formed farther from the Sun retained large quantities of hydrogen, either in the form of molecular hydrogen or in compounds such as ammonia and methane. A very large planet (Jupiter) and three other planets of smaller size, but still much larger than the terrestrial planets (Saturn, Uranus, and Neptune), also formed in the Solar System.

A nonnegligible fraction of the planetesimals did not merge into large planets but formed an enormous number of small bodies, from the satellites of the large planets to asteroids and comets. The small bodies that formed in the outer, and therefore colder, parts of the cloud were, and still are, very rich in volatile substances in solid form (ice). In particular, many of them are rich in water ice.

No planet in our system grew to a mass that, during gravitational collapse, would allow it to attain temperatures high enough to start thermonuclear reactions and therefore to become a star. If Jupiter had grown larger (much larger, by more than an order of magnitude), the Sun would have been a double star, like so many others in our galaxy.

Something actually exists that is a halfway stage between a large planet and a star: objects with mass equal to some tens the mass of Jupiter, but smaller than one-tenth the mass of the Sun, do not succeed in starting enough thermonuclear reactions to become actual stars, but they do produce heat and therefore radiate strongly in the infrared region of the electromagnetic spectrum. These objects are usually defined as *brown dwarfs*.

The gravitational field in the Solar System is extremely complex, and therefore the orbits that would be elliptical in the case of the "two bodies problem" (i.e., if only the Sun and one planet were present) are perturbed by the other planets, and particularly by Jupiter. The presence of Jupiter had a great influence on the orbits of the planetesimals and particularly of those that orbited close to the giant planet. The reason why no fully grown planet, but only many small bodies, exists today in the zone between Mars and Jupiter (the asteroid belt) is thought to be that the presence of such a large body prevented the formation of a planet. The theory that a planet was present there but later exploded, giving origin to the asteroids, is no longer considered viable. Many interpretations of the history of the Solar System, and of life on Earth, that still have such a theory as their basis can be defined as whimsical.[6]

Beyond Neptune's orbit, in the outer part of the Solar System, many small bodies formed, constituting the so-called Kuiper belt; the largest of such bodies are Pluto and its satellite Charon. The average radius of Pluto's orbit is 5.9 billion km (39.6 astronomic units, AU[7]). Since the orbit of Pluto is strongly elliptical and crosses the orbit of Neptune, sometimes the latter (with an average radius of 4.49 billion km, i.e., 30 AU) is assumed as the external limit of the Solar System. A schematic map of the Solar System is shown in Figure 3.1; the orbits are drawn to scale, while the sketches of the planets are much enlarged. The main data of the planets and the Moon are listed in Table 3.2.

Farther out is an enormous number, probably tens or hundreds of billions, of blocks of ice of all sizes that remained in very wide orbits and constitute the so-called Oort cloud. Obviously, such a cloud has not been observed experimentally, since it is not possible to see such small and distant objects even with the most powerful telescopes, but its existence

---

[6] T. Van Flandern, *Dark Matter, Missing Planets & New Comets*, North Atlantic Books, Berkeley, 1993.

[7] One astronomical unit (AU) is equal to the average distance of the Earth from the Sun, 149.47 million km.

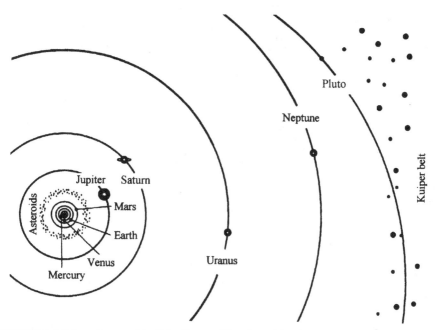

FIGURE 3.1 Schematic map of the Solar System. The orbits of the planets are plotted to scale, while the sketches of the planets are enlarged by a factor of 1000.

and characteristics have been deduced from the orbits of the comets, which are elements of the cloud whose orbits have been perturbed by stars passing close to the Solar System and made to enter its inner zone. The Oort cloud contains an enormous number of comet nuclei, ready to enter the inner Solar System where, because of the heat, their external layers evaporate and are pushed by the pressure of light and by the solar wind to form the spectacular tail.

The intermediate zone of the Oort cloud is expected to be at a distance of about half a light-year[8] (around 32,000 AU) from the Sun. From what we know at present, it is a kind of spherical shell surrounding our system and starting, according to the theories, well beyond the orbit of Neptune (radius about 30 AU), at roughly 10,000 to 40,000 AU from the Sun. The zone of maximum density would be between 30,000 and 60,000 AU and its extreme limit at 100,000 AU, almost two light-years from the Sun. If

[8] One light-year, the distance light covers in a year, is equal to 9461 billion km or about 63,300 AU. Another unit often used for stellar distances is the parsec (ps, the distance at which the orbit of the Earth is seen under an angle of a second of degree); 1 ps is equal to about 3.2 light-years. For very great distances, kps or Mps, equal to 1000 ps or 1,000,000 ps, respectively, is sometimes used.

TABLE 3.2 *Main data of the planets of the Solar System and the Moon. The diameter* D *is in km, the mass is referred to the mass of the Earth* M/M_e *(*$M_e = 5.96 \times 10^{24}$ *kg), the density* ρ *is in* kg/m^3, *the average radius of the orbit* R_0 *is in millions of km, the orbital period* T *in years (in days for the Moon), the gravitational acceleration at the surface* g_0 *is in* m/s^2, *and the escape velocity* V_e *in km/s.*

	D (km)	M/M_e	ρ (kg/m^3)	R_0 (Gm)	T (years)	g_0 (m/s^2)	V_e (km/s)
Sun	1,320,000	332,000	1,410	–	–	274	616
Mercury	4,879	0.0543	4,400	58	0.241	3.2	3.95
Venus	12,104	0.813	5,000	107.6	0.615	8.64	10.28
Earth	12,756	1	5,520	149.3	1	9.81	11.18
Mars	6,794	0.107	4,100	228	1.881	3.95	5.13
Jupiter	142,984	318.4	1,350	777	11.86	26	60.4
Saturn	120,536	94	710	1,422	29.46	11.48	36.64
Uranus	51,118	14.5	1,250	2,860	84.01	10.32	22.16
Neptune	49,520	17.2	1,600	4,490	164.8	14.5	25.6
Pluto	2,320			5,900	247.7		
Moon	3,400	0.012	3,300	0.384	27.3 (days)	1.583	1.583

the nearest stars also have a similar cloud of comet nuclei, the peripheral zones of the Oort cloud would be almost in contact.

In the initial phases of the formation of the rocky planets, when impacts were extremely frequent, their temperature was very high and the organic molecules present in the inner zone of the protoplanetary nebula were, for the most part, decomposed. Hydrogen and the other light elements that were present in organic substances and in water (provided that some water was still present in the inner part of the protoplanetary nebula) got away from the planet and were lost in space. In the initial phase of their history, the terrestrial planets were therefore extremely poor in organic substances, water, and light elements in general.

When the initial bombardment started to slow down, because most planetesimals had already been swept up by the planets in their formation, the temperature of the Earth and the other rocky planets began to decrease and water brought by the comets—which continued, though in a more sporadic way, to fall on them—remained on the planet. Depending on the temperature, it was in solid, liquid, or vapor form. The organic compounds brought by meteorites and comets also remained on the surface of the planets.

Through this stage of the formation of the Solar System, chemical evolution continued in comets, meteorites, and on the surface of rocky planets. The variety of organic compounds that have been found in comets

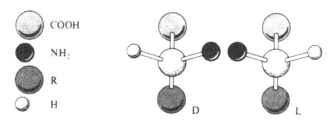

FIGURE 3.2 *Molecular structure of the two forms (D and L) of the same amino acid. NH₂: aminic group; COOH: carboxylic group; R: radical (if, for instance, R=CH₃ the amino acid is alanine). The central white sphere is the carbon atom.*

and meteorites is remarkable and includes very complex molecules. Among the various types of meteorites that have been identified, the carbonaceous condrites, whose composition is considered typical of the main belt asteroids (the asteroids that formed between the orbits of Mars and Jupiter) contain large quantities of carbon and its compounds. For instance, the study of the Murchison meteorite, which fell in Australia in 1969, led to the identification of organic substances of various kind, including 74 different amino acids. Among the amino acids that have been identified are eight of the 20 that constitute the proteins from which living matter is made, and another 11 that exist in the cells of terrestrial organisms. The remaining 55 do not play any role in terrestrial biology and do not exist on our planet; they therefore constitute convincing evidence that the presence of amino acids in the meteorite is not due to contamination subsequent to the fall. Besides, the contamination is also unlikely owing to the fact that most of the almost 80 kg of material of the Murchison meteorite was recovered a short time after the impact.

It must be noted that every amino acid can exist in two different forms: dextrorotatory (D) and levorotatory (L)—terms that come from the ability of a solution of amino acids in water to rotate the plane of polarization of light crossing the solution[9] toward the right or the left. This property of many organic substances is defined as *chirality* and is due to its molecular structure: the two molecular forms, though having the same composition and chemical properties, have specular structures (Figure 3.2).

All the amino acids that constitute the proteins synthesized by living beings on Earth and almost all natural amino acids are L, while sugars are D. On the other hand, if amino acids and sugars are synthesized through

---

[9] Actually the distinction is more complex, since the direction of rotation of the plane of polarization of the light also depends on other factors, and the distinction between the D and L forms of amino acids depends on the configuration taken by the carbon atom (see Figure 3.2).

nonbiological reactions, molecules of type D and L are obtained with equal probability. Note that only L amino acids are of interest in alimentary or pharmaceutical applications, since D amino acids are not biologically active.

The amino acids found in meteorites are of both L and D type, which provides more evidence that they are not the result of terrestrial contamination. Until a few years ago the two forms were thought to be equally present, while recent studies of the Murchison meteorite seem to show a certain prevalence of the L form, even if very weak.

The presence of only one of the two forms in living beings could be due to chance. This would be strong evidence that all living beings on Earth descend from a single ancestor and if this were the case, finding a living being containing amino acids of the "wrong" type would be proof of its extraterrestrial origin. It could, on the other hand, be due to the fact that only the L type is suitable for life, but this a questionable hypothesis indeed, since the chemical characteristics of the two forms are identical. The truth is, we do not know.

Recently the hypothesis has been advanced that the Solar System, in its early stage, passed close to a neutron star, which could have caused a certain preference for L type amino acids that would therefore be the most common form in the whole Solar System.

It must be remembered that structures identified by some scientists as microfossils were found in some carbonaceous condrites, particularly in the Murchison meteorite. This finding, which is not endorsed by the majority of scientists, has given new impulse to the theories according to which life came to Earth from space (panspermia, see below).

A doubt must be dissipated here. How is it possible to find complex organic molecules, or even life-forms, in a meteorite that had to cross the atmosphere at very high speed as it fell to Earth? Doesn't aerodynamic heating, which destroys the smallest meteorites completely, producing the typical bright trail, completely destroy all organic matter?

The answer is that the intense heat affects only a part of the surface of the meteorite, which suffers strong heating and a process of vaporization (ablation), like the heat shields of space vehicles. The inner parts of the meteorite, however, protected by the outer layer and cooled by the vaporization of the latter, remain relatively cool and can undergo a rise of temperature limited to a few degrees. In these conditions a meteorite works like a reentry capsule and the substances it contains can land intact.

Even though meteorites, such as the Murchison, are very spectacular and attract much attention, by far the largest part of the cosmic matter reaching the Earth is in the form of micrometeoroids, with a size of fractions of a micrometer (one thousandth of a millimeter).

It has been calculated that since the formation of our planet, micrometeoroids brought on its surface a quantity of carbon equal to 3×10^{13} (30,000 billion) tons, more than 30 times the total mass of carbon present in Earth's biosphere (about 10^{12} tons).

Earth contained neither water nor carbon at its formation, but it quickly became rich both in water (brought in mainly by cometary nuclei) and organic substances (brought in by cometary nuclei, meteorites, and micrometeorites). As soon as the surface temperature decreased enough to allow the existence of liquid water, conditions were present for the continuation of the chemical evolution that would eventually result in the beginning of life.

THE FORMATION OF EXTRASOLAR PLANETS

The theory of the origin of the Solar System with the largest following today is that the Sun and the planets originated from the condensation of a protoplanetary nebula. Alternative theories—which essentially assume that the planets formed later than the Sun, following a catastrophic event involving our star—although once considered more likely, have now almost disappeared.

Since all stars are believed to have been formed in a similar way, the existence of planetary systems around other stars should be the norm and not the exception, as many believed just a few years ago. If, for instance, the planets had been caused by the gravitational attraction of a star that passed very close to the Sun, the extreme improbability of such an event would make the existence of a planetary system a rare case indeed.

As already mentioned, the formation of planets is probably linked to the angular momentum of the protoplanetary nebula: if it were very small, the whole mass of the cloud would collapse into the central star. This condition appears to be extremely unlikely, and at any rate it was not the case for the Solar System: while the largest part of the mass is concentrated in the Sun, the angular momentum is, so to speak, almost all stored in the planets and in the smaller bodies.

In the same way, when a double star is formed there is no need for the existence of planets to absorb the angular momentum of the nebula (this is true only if both stars come from the same nebula and are not two celestial bodies that originated separately and were later gravitationally captured by each other). Furthermore, the presence of a second large body could make it impossible for planets to form in orbit around double or multiple stars, just as the presence of Jupiter prevented the formation of a planet between Mars and Jupiter. The possibility of the existence of planets in multiple star systems is an open problem, and it is likely that planets do not exist in the rare case of multiple stars in which the two main bodies are very close. However, if the two stars are at a great enough distance from each other, stable orbits exist around each of the components of the system and around both. In the latter case the orbit should be extremely wide.

In the case of Alpha Centauri, for instance, the separation of the two main components is 23 AU (a little more than the radius of the orbit of Uranus, 19 AU). It is not clear whether in these conditions stable orbits around each component can exist, but if a rocky planet existed on an orbit similar to that of the Earth around the main component—a star very similar to the Sun—the secondary component would be seen from the planet like a very bright star. The third component is extremely far away and very small, and therefore should not create problems.

Two pieces of evidence are normally adduced for sustaining that many stars have a planetary system: protoplanetary nebulas have been observed around stars in formation and, since the time observational techniques have discovered planets in orbit around other stars (the first discovery, by Mayor and Queloz, occurred in 1995), new planets have been found at such a rate that 146 planetary systems had been observed by the end of 2005, 18 of which were multiple, for a total of 170 extrasolar planets.[10]

The existence of planets around other stars had been hypothesized since the time it was discovered that the Sun is nothing other than a star, more or less like the others. Some astronomers in the past thought they had discovered extrasolar planets, but the discovery of an object of planetary size around a star of solar type is extremely difficult and is still at the limits of modern instrumentation. Direct observation with optical telescopes is not possible: the weak light reflected by the planet is completely hidden by the light of its star. Only when the planet passes between the star and the observer is it possible to notice a decrease in the brightness of the star, but the event is rare and conditioned by the position of the orbital plane. Above all, the effect is so small that extremely accurate observations are required. Interferometric studies and other observational techniques performed with present-day instruments are insufficient to reveal extrasolar planets, so instruments designed specifically for this task are now being studied.

The only way of identifying objects, like planets, that do not shine in their own light is to study the perturbations they cause in the motion of

---

[10] *Extrasolar Planets Catalog*, http://vo.obspm.fr/exoplanets/encyclo/index.php.

other known celestial bodies. There is nothing new in this: the planet Neptune, unknown in ancient times, was discovered by studying the anomalies of the orbits of other planets that could not be explained by the presence of known bodies. To explain these anomalies, the presence of an unknown planet in a certain position was assumed and, when the telescopes were aimed toward that point, a planet was indeed found.

This method therefore works, and for the outer planets of the Solar System it was easy enough to apply it with the observational techniques of the past. When one wants to search for planets in orbit around other stars, however, the difficulties become overwhelming, and extremely accurate observations must be performed over periods of years.

An example of the difficulties in the search for extrasolar planets is the controversial case of Barnard's Star system, whose discovery by E.E. Barnard was published in 1916 in *The Astronomical Journal* and in *Nature*. It is a rather small and dim star that has a notable characteristic: it is the star with the highest proper motion ever discovered, about 10.3 arcseconds per year. It is a red dwarf, at only 1.82 parsecs (5.95 light-years) from the Sun.

The astronomer Peter van de Kamp, of the Sproul Observatory, began a search that kept him busy for his entire life, during which he took as many as 2000 photographic plates of the star from 1938 to 1962. The motion of the star made him think it had a planet, and he calculated that its mass had to be equal to about 1.6 times the mass of Jupiter and that its orbital period had to be 24 years. These data were refined using plates taken in the period 1916–1919, and finally, in subsequent papers, van de Kamp announced the discovery of a second planet around that star: the periods of the two planets were 26 and 12 years and the masses equal to 1.1 and 0.8 the mass of Jupiter. In 1975 the evaluation of the mass was corrected to 0.4 and 1.0 the mass of Jupiter and in 1982, the same van de Kamp published further corrections.

Other astronomers, such as Gatewood and Eichorn, performed measurements on plates taken using different telescopes, but did not notice any motion of the star to indicate the presence of planets. Hershey performed measurements on other stars using the same plates that van de Kamp had used and noticed similar motions for other stars, too. The hypothesis was thus formulated that the apparent motions of Barnard's Star were due not to the presence of planets but to optical errors of the telescope originally used. Van de Kamp continued until his death, in 1995, to claim that the planets he discovered actually existed, but few other astronomers supported his statement. After more than half a century of research, the existence of planets around Barnard's Star, one of the closest to us, is more uncertain than ever.

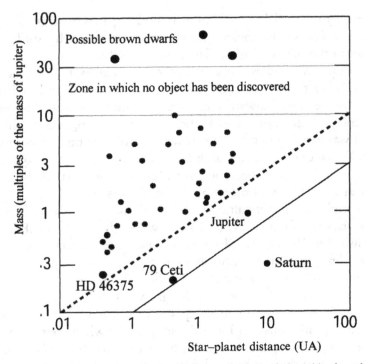

FIGURE 3.3 Mass and distance from the star of the extrasolar planets discovered by the end of 2000. For comparison, Jupiter and Saturn are also reported; the dashed and the continuous line indicate the respective limits of observation with different techniques.

Recently, however, the search for extrasolar planets was pursued by measuring the speed of a star using the Doppler effect, rather than measuring its position. With the present instruments it is possible to identify only very massive planets orbiting at a short distance from the star; if there were a system similar to the Solar System, it would be impossible to identify the Earth, and even Jupiter would be at the limits of observational possibilities. The limits of observability with the present instruments are plotted in Figure 3.3, together with the mass and the distance from the star of the planets actually discovered by the end of the year 2000. The mass is expressed in multiples of the mass of Jupiter, which is equal to 318 times the mass of the Earth.

Moreover, the possibility of discovering a system depends on the orientation of the plane in which the orbits lie: if the plane is perpendicular to the line of sight, it is not possible to identify the planets. The mass of the planet measured in this way depends on the angle i between the plane of the orbit and the line of observation: the measured value is therefore not the mass m of the planet but the product $m \times \sin(i)$. Since the sine is always a number smaller than 1, the measured value is a lower limit to the actual value. Starting from this consideration, some astronomers think that the discovered planets are not really planets but small stars (brown dwarfs) in much inclined orbits. This objection, however, is usually rejected on a statistical basis (it is impossible that *all* the observed systems should have such unfavorable orientations of the orbital planes) and the existence of extrasolar planets is given for sure by the majority of astronomers. Apart from the planets orbiting normal stars, some planets orbiting pulsars were also found.

Ten years after the first confirmed sighting of an extrasolar planet, it is possible to affirm not only that planets exist outside the Solar System, but also that the number of stars having planets is certainly very large. The percentage of stars having a very regular planetary system like ours, in which the orbits of the planets are stable and almost circular, is unknown. Also unknown is whether a wide variety of bodies exists in those systems: giant planets mostly made up of gas, rocky planets of a terrestrial type, smaller bodies orbiting around the larger planets, and finally a crowd of asteroids and comets.

The main characteristics of the 22 stars nearest to us are reported in Table 3.3 and their position with respect to the Sun is reported in the three-dimensional plot of Figure 3.4. Most of them are quite dim stars (their intrinsic brightness is very low, and therefore, even if they are very close to us, astronomically speaking, they appear inconspicuous to an observer on Earth), but among them are some very bright and well-known stars, like Alpha Centauri, Sirius, and Procyon. Six of them, including Sirius and Procyon, are double stars and one, Alpha Centauri, is triple.

One of them, Epsilon Eridani—a star similar to the Sun even if a little smaller and colder—has a planet with an apparent mass of 0.86 the mass of Jupiter, with a larger half-axis of the orbit of 3.3 AU. If it were not for the high eccentricity of the orbit, this could be a sign of a planetary system similar to ours. Another star, Lalande 21185, probably also has a planet, but its discovery has not yet been confirmed.

The plot of Figure 3.4 is centered in the Sun, with the z-axis parallel to the axis of rotation of Earth in 1950 (the positive direction is pointing north) and the x-axis parallel to the intersection between the equatorial plane of Earth and the ecliptic (plane of the orbit of Earth).

The process of formation of the planets seems to have been much determined by chance; the collisions between the planetesimals can give way to bodies with very eccentric elliptical orbits, which, if massive, can perturb the orbits of the other planets causing such instabilities that some planets could fall into the star (in the Solar System many comets, whose

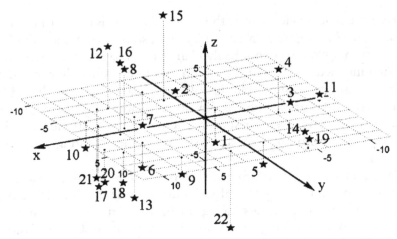

FIGURE 3.4 *The 22 stars closest to the Sun, represented in a three-dimensional plot. The numbers are referred to in Table 3.3; the scales are in light-years.*

TABLE 3.3. *The 22 stars nearest to the Sun, in order of distance*

No. Star		Dist. (l.y.)		Spectral class	Luminosity (Sun=1)	Mass (Sun=1)
1	Alfa Centauri	4.4	Triple	G2–K6–M5e	1.3–0.36–0.00006	1.1– 0.89– 0.1
2	Barnard's Star	5.9		M5	0.00044	0.15
3	Wolf 359	7.6		M83	0.00002	0.20
4	Lalande 21185	8.1		M2	0.0052	0.35
5	Sirius	8.7	Double	A1–DA	23.0–0.0028	2.31–0.98
6	Luyten 726-8/UV Ceti	8.9	Double	M6e–M6e	0.00006–0.00004	0.12–0.10
7	Ross 154	9.5		M5e	0.0004	0.31
8	Ross 248	10.3		M6e	0.00011	0.25
9	Epsilon Eridani	10.7		K2	0.30	0.8
10	Luyten 789-6	10.8		M6	0.00012	0.25
11	Ross 128	10.8		M5	0.00033	0.31
12	61 Cygni	11.2	Double	K5–K6	0.063–0.040	0.59–0.50
13	Epsilon Indi	11.2		K5	0.13	0.71
14	Procyon	11.4	Double	F5–DF	7.6–0.0005	1.77–0.63
15	+ 59° 1915; 173739	11.5	Double	M4–M5	0.0028–0.0013	0.4–0.4
16	Groombridge 34	11.6	Double	M2–M4	0.0058–0.0004	0.38–...
17	Lacaille 9352	11.7		M2	0.012	0.47
18	Tau Ceti	11.9		G8	0.44	0.82
19	Luyten BD + 5° 1668	12.2		M4	0.0014	0.38
20	L725-32	12.5		M5e		
21	Lacaille 8760	12.5		M1	0.025	0.54
22	Kapteyn's Star	12.7		M0	0.004	0.44

orbit was perturbed by the largest planets, mainly by Jupiter, have been observed to fall into the Sun) or be expelled into interstellar space.

Some of the recently discovered extrasolar planets have a mass greater than that of Jupiter and very elliptical orbits: the existence of smaller planets on stable and, above all, circular orbits in such systems is questionable.

Other planets have masses of the order of that of Jupiter and are very close to the star, their orbit being much smaller than the orbit of Mercury around the Sun. This situation is defined as a "planet of the 51 Pegasi type," from the name of the star that is orbited by the first planet of this kind to be discovered. The radius of Mercury's orbit is 0.4 AU, that of the planet orbiting around 51 Pegasi is 0.05 AU, and planets with even smaller orbits have been found. It is not clear whether massive planets can form on such small orbits, and the most common explanation is that the planet originally formed at a greater distance from the star, then moved inward on a spiral trajectory. If this is true, a planet of this type would have completely cleaned up a large zone of the planetesimals or previously developed planets, to the point that only one planet would now exist in the system.

There may be very few systems containing rocky planets at a distance from the star suitable for life to develop. On the other hand, large planets close to the star or on elliptical orbits are in the best conditions to be observed, and therefore the cases mentioned above could be rare and anomalous occurrences that appear to be more common than they actually are.

Besides, terrestrial-type planets could form as satellites of large gas planets in an orbit close to the star, or form independently and then be captured by a gas planet that got close to the star. An alternative way of surviving on the orbit of a large planet, other than gravitational capture, can be by being captured in one of the Lagrange points on the orbit of the planet. In the Solar System this position (on the orbit of Jupiter, 60° leading and 60° trailing the planet) is occupied by two groups of asteroids, named after Greek and Trojan heroes. For this reason planets or asteroids in the Lagrange points on the orbit of a large planet are said to be "Trojans."

Even if many planetary systems contain no habitable terrestrial-type planets, they might have habitable satellites or Trojans. They would, however, be extremely difficult to discover from a great distance.

Chapter 3

THE BIRTH OF LIFE ON EARTH

For now, the only planet we know that can host life is Earth. Every bioastronomic study must therefore begin from this one case.

It is very difficult to define what life is. If one compares a higher animal, for instance a cat, with an inanimate object like a stone, the difference is intuitive and there seems to be no need to resort to precise definitions. But with simpler forms of terrestrial life the distinction is much less clear, to the point that there have been arguments on whether viruses should be considered living beings.

Life is often defined as the ability of an organism to grow and reproduce, but such a definition can also be applied to chemical systems that are not usually considered to be "living."

Often the ability to evolve through random events is added to these two characteristics. The ability to grow and reproduce implies the organization of matter from the external environment according to a well-defined structure, something causing a decrease of entropy in a limited portion of space. To do this, the living being needs to extract energy from the environment and therefore to exploit conditions of thermodynamic disequilibrium. But this last characteristic, essential though it is for life, cannot be assumed as the discriminating factor distinguishing a living being from an inanimate object, since thermal engines, those of cars for instance, work in the same way. Yet another basic characteristics of life is its ability to store information. One of the many possible definitions of life is therefore as follows: *Life is a self-sustained chemical system, capable of undergoing Darwinian evolution.*

When speaking of life, we often add "as we know it" or "of terrestrial type," intending by this to refer only to living beings, based on the chemistry of carbon, that store the information needed for their growth and reproduction in the complex structure of the deoxyribonucleic acid (DNA). Whether forms of life based on other elements (for instance, on silicon) may exist and whether living beings can use information storage mechanisms of a completely different kind is a question that has not yet been answered. If this were the case, it could be very difficult to recognize such organisms as living beings.

The preceding sections described how the Earth was formed, about 4.5 billion years ago, through aggregation of planetesimals. During the first 500 million years of its history it sustained a very heavy bombardment by asteroids and comets—a bombardment that slowed down as soon as the gravitational attraction of the planets had cleaned up the Solar System from these residuals of the nebula.

Toward the end of this growth process, Earth received a terrible blow: a traumatic event that gave origin to the Moon. For a long time many different theories on the origin of the Moon thrived, but all were swept away by the scientific results of the explorations performed by the astronauts of the *Apollo* missions and by the laboratory analyses of the specimens they brought back to Earth. Today the "origin of the Moon" theory that has the largest credit is that it was torn from Earth when a large celestial body, with a mass equal to that of Mars or even greater, hit Earth when its process of formation was almost completed. The nuclei of the two celestial bodies fused together in the impact and a part of their mantles, mainly made of silicates, was expelled in space, remaining in orbit around the planet. The large satellite solidified in an orbit much lower than that presently occupied by the Moon (with a radius of about 24,000 km, instead of the present 386,000). The duration of the month was shorter than at present, while the speed of rotation of the planet around its axis was much higher (a day lasted about 5 hours).

The interaction between a planet and its satellite is rather complex. If the mass of the satellite is large and the radius of its orbit is small, the planet and the satellite deform, as shown in Figure 3.5. In the case of Earth, the tides cause fairly large variations of the level of the oceans, but they also deform the crust of the planet. Because of the energy dissipations accompanying these deformations, the bulges of the surface of the planet are not directed exactly along the line connecting the centers of the two bodies. The result of this dissipation of energy is, in the first place, a deceleration of the speed of rotation of both bodies around their axes, to the point that they may end up always showing the same side to each other (synchronous rotation). The Moon is already locked in this situation, while the rotation of Earth continues to slow down.

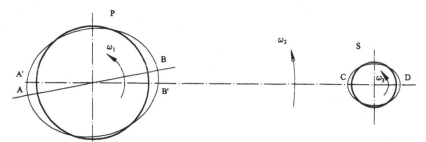

FIGURE 3.5 *Deformation of the planet P and of the satellite S due to tides. The satellite rotates at a speed ω_3, equal to the speed ω_2 of the motion of revolution (it is gravitationally locked owing to tidal forces) and therefore the swellings C and D are lined up with the line connecting the centers of the two bodies. The planet rotates more quickly $\omega_1 > \omega_2$, and therefore swellings A and B are not located in A' and B'. This effect causes the deceleration of the rotation of the planet.*

Besides, because of energy dissipations inside the mass of the planet, an energy transfer occurs between the rotational motion of Earth and the Moon and the revolution of the Moon around Earth, with the consequence that the first actions decelerate and the distance of the satellite gradually increases. Today, the Moon still moves outward by about 4 cm per year.

These tidal effects are very common and bring many satellites to lock in synchronous rotation and move away from the planet, if the rotation of the planet has the same direction as the revolution of the satellite. If the motion of the satellite is retrograde, as in the case of Triton, a satellite of Neptune, the orbit lowers until the satellite falls onto the planet.

Apparently, this explanation of the origin of the Moon makes the case of a satellite of size not much smaller than that of the planet around which it orbits—already atypical in the Solar System—a rare event in general.

For a satellite to form in this way, not only must a large body hit an almost completely formed planet—an exceptional event in itself—but it must also hit the planet at the correct speed and with the correct angle of impact. In general, rocky planets in our system do not have satellites (Mercury, Venus) or have small satellites that are clearly captured asteroids (Mars, which is close to the asteroid belt). It is probable that Earth would also have had no satellites, were it not for an exceptional accident. This consideration may be important, as we will see later, for the possibility of the existence of extraterrestrial life. If the Moon played an important role in the development of life, life would predictably be very rare.

Recently, however, it has been suggested that the body that impacted Earth was formed in one of the Lagrange points on Earth's orbit, and some computer simulations have shown that in this case the complex dynamics of the Solar System would inevitably cause it to hit Earth with an energy suited to creating a large satellite. If this is true, the formation of the Moon was not a random event, and consequently the Earth–Moon system would not be an oddity. At any rate, the only other case of a small planet with a large satellite, Pluto, is entirely different, since both are thought to be planetoids of the Kuiper belt.

About 4 billion years ago, the bombardment of asteroids and comets on a not yet fully formed Earth and on the Moon was much less heavy, even if it was far stronger than at present. From the clues found on the Moon by the *Apollo* astronauts, a date of 3.85 billion years ago was obtained for the end of the meteoric bombardment—a date that is now universally accepted.

It is very difficult to assess when life started on our planet. Many think that life appeared on Earth in a very short period of time: the most ancient fossils of which we have evidence are 3.5 billion years old, while more ancient, even if less certain (although likely), clues of living beings are dated at 3.65–3.7 billion years ago. They essentially consist of granules of graphite of probable biological origin; from these it would result that life began on our planet no more than 200 million years after the slowing down of the meteoric impacts made it possible.

However, there are scientists who dispute these statements; they date the most ancient evidence of life at 2.1 billion years ago, questioning the possibility that life formed a short time after the formation of the planet. It is also possible that life started even earlier than 3.7 billion years ago, only to be wiped out completely or suffer many hard blows by traumatic events like asteroid or comet impacts.

Life is, so to speak, very clever in wiping out its own traces and, above all, it is understandable that nothing should remain of the first beings, living before the end of the process of formation of the planet. It is therefore extremely unlikely that we can ever find fossil remains of the first living beings that inhabited our planet.

Recently a new possibility of discovering traces of very ancient life has been suggested: during the early times of Earth's evolution, owing to the heavy bombardment, many fragments of the planet were sent into space by the impacts. A good number of fragments could have fallen on the Moon, which at that time was much closer than now. If at that time there was life, traces of ancient life-forms could be available in these fragments of our planet, which are now scattered on the lunar surface. Owing to the lack of biological activity and the scanty geological activity, they should be much better preserved than those that remained on our planet. So, it seems rather odd that to find traces of early terrestrial life we should need to go to the Moon!

Another problem is the difficulty of recognizing traces of ancient life (if we can find them), particularly if life started on our planet more than once, each time based on different patterns. The first living beings whose traces we found were extremely simple: not only were they unicellular organisms, they were prokaryotes, that is, cells without a nucleus, in which the material containing the genetic information is directly stored in an undifferentiated cytoplasm. Yet simple as they may be, the gap between these forms of life and the most complex organic molecules found in interstellar clouds and meteorites, like amino acids, or those obtained from simple chemical substances in various laboratory experiments, is still enormous.

Some biologists, such as the Nobel laureate Jacques Monod, author of *Chance and Necessity*, asserted that life is due to chance, and therefore such an unlikely anomaly makes Earth something really unique in the Universe.

It has been calculated that the probability of producing a living organism by combining together at random some amino acids is of the order of 1 divided by 10^{130}, a number corresponding to 1 followed by 130 zeros. The number of stars in the visible Universe has been evaluated as 10^{20}. Comparing the two numbers it is easily deduced that the probability that living beings exist is unimaginably small. Calculations of this kind are undoubtedly somewhat arbitrary, but a probability of this kind is unbelievably smaller than the proverbial chance of a monkey, typing at random on a keyboard, producing a meaningful book.

Since we know for sure that such an event occurred once, with such a low probability it is almost certain that it cannot be repeated. Monod would then be right in asserting that humans are alone in the Universe, and well aware of it. But to state that such an unlikely phenomenon occurred by chance is very unsatisfactory and even more so today, since we now know that only a short time (on a cosmic scale even a billion years is not a very long time, compared with the infinitesimal probability mentioned above) passed from the instant when life was given the possibility to start on Earth and when it actually started. This seems to show that the probability that life developed from nonliving matter is high.

It therefore looks as if the development of life is somehow a direct consequence of certain laws of physics causing matter to organize in structures of increasing complexity, up to living organisms. From all this a conclusion may be drawn: either life on Earth is unique, or life is extremely widespread in the Universe; it is unlikely that a third option exists.

However it started, life since its beginning had a strong tendency to differentiate. Until some decades ago it was customary to divide all forms of life into two groups, the vegetable and animal kingdoms, but today it is clear that the picture is far more complex: a first division is between *prokaryotes*, beings (usually unicellular) made of cells without a nucleus, and *eukaryotes*, whose cellular structure is more complex. The prokaryotes are divided into two classes: the *archeobacteria* (or archea) and the *bacteria* (eubacteria). The archeobacteria and the bacteria seem to have common ancestors, probably the first living things on Earth.

Figure 3.6 presents a *tree of life* (a kind of genealogical tree of the living species). In the lower part, marked by a question mark, the hypothetical path that could have led from nonliving matter to the prokaryotes, archeobacteria and bacteria, is shown. Today we know almost nothing of a development that could have implied a complex

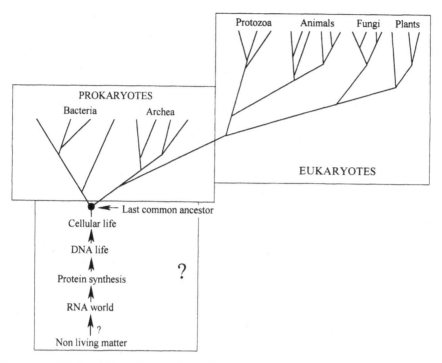

FIGURE 3.6 *A possible and extremely simplified "tree of life," a kind of genealogical tree of the living species. In the lower part, marked by a question mark, the hypothetical path that could lead from nonliving matter to the prokaryotes, archea and bacteria, is shown.*

branching of genera and species that later became extinct, or it could include various *trees of life*, which started when the meteoric bombardment was still very strong and, subsequently, were completely and traumatically annihilated.

All present forms of life are based on a cellular structure and the formation of proteins, with the exception of viruses, being performed by using genetic information contained in DNA. Yet the latter can only exist in the presence of proteins: we have no idea how this chain might have begun. Besides DNA, all forms of life use ribonucleic acid (RNA) to transcribe and translate information stored in the DNA.

Viruses are extremely simple beings, without a cellular structure, on the border separating nonliving matter from life. They are very small organic macromolecular systems, from 20 to 300 nm (nm = one millionth of a millimeter), visible only using electron microscopes, made by a protein membrane containing one nucleic acid only, either DNA or RNA. Viruses can only reproduce if they enter a cell and use the mechanisms of cellular synthesis of the host: it may be said that they do not have life outside a host cell. The simpler forms of life that existed before the prokaryotes could not be similar to viruses, whose origin is usually ascribed to sections of cellular DNA or RNA which became self-sufficient, or to a simplification of cells that lost their enzymes. It seems then to be a kind of involution—intending by the term a simplification of a more complex structure—that occurred after life had already developed.

Currently life is thought to have started with forms that preceded DNA, and that once life based on DNA spread, these primitive forms disappeared. The first living forms would have used RNA instead of DNA to store their genetic information: so an RNA world that would have been a forerunner to the DNA world has been assumed (Figure 3.6).

Archeobacteria are generally extremophile beings; that is, they are suited to living in conditions that we consider extreme. For instance, there exist thermophile archeobacteria that live at temperatures up to almost 100°C, temperatures in which no eukaryote could live; hyperthermophiles that live in extremely warm environments (some archeobacteria can withstand temperatures over 200°C); and acidophiles that live in extremely acidic environments; and so on.

Recently some archeobacteria and bacteria able to live underground have been discovered. Samples brought to the surface from depths of more than 3000 meters while digging wells were found to be rich in forms of life. These forms of underground life are so numerous and widespread as to suggest that the total mass of underground living matter is greater than that existing on the surface. The image of life like a thin and fragile shell on the surface of the planet is therefore probably wrong.

Hyperthermophile archeobacteria living close to hydrothermal vents deep in the oceans are particularly interesting from the bioastronomical point of view. They are very simple organisms, able to withstand temperatures higher than 150°C, which exploit the temperature gradients existing in the proximity of the hot vents for their energy needs.

Living beings may use organic material that has already been "processed" by other organisms, like animals that feed on other animals or vegetables, or they can use inorganic material and energy taken from the environment, like plants that start from inorganic material from the ground and the atmosphere and, through the photosynthesis reaction, use the light of the Sun to synthesize proteins and sugars. The latter beings (autotrophs) are therefore at the beginning of any food chain.

In the past, when all autotrophs living on Earth were thought to use photosynthesis, it followed that all the energy sustaining the terrestrial biosphere ultimately came from the Sun. But the discovery of thermophile and hyperthermophile archeobacteria showed that autotroph beings can draw the energy to sustain life directly from the thermal energy that originates inside the planet, to a large extent because of radioactive decay. Other archeobacteria living at great depth in the ground draw their energy directly from the chemical energy of substances they absorb from the ground. This discovery is important, because it shows that the presence of light is not essential for life and that radioactive decay may sustain life at great distances from any star or deep under the surface of a planet.

There are many clues that the common ancestors of bacteria and archeobacteria, from which all living beings on our planet descend, were of that type, and that life on Earth began independently of solar energy. Living beings whose habitat is in the depths of the ground or under thousands of meters of ocean are also more protected from catastrophes caused by meteorite impacts, so it is reasonable to assume that life started in protected places and only later, when the impacts thinned out, migrated to shallower waters.

This hypothesis contradicts what was considered to be a scientific fact until a few years ago, namely that life started either close to the ocean surface or in tidal pools on the coasts, which, because of the strong tides due to the Moon being much closer to the Earth than today, were periodically awash in currents that mixed waters and increased the quantity of gas dissolved in it. Tidal pools were also subjected to the electric phenomena that many think were essential in the synthesis of the first organic materials. Most scientists believed that life originated in such tidal pools, and this suggested that the presence of a large satellite, causing strong tides, is essential for life to start.

The water in the tidal pools was probably rich in organic substances and had a high enough temperature to speed up chemical reactions, thus allowing nature to perform a wide range of experiments. This was how Charles Darwin imagined the origin of life: he spoke of a shallow pond, heated by the Sun. Chemical evolution would complete its cycle there and, almost imperceptibly, what many called the *primordial soup* would favor the shift from inanimate matter to living beings.

In a famous experiment performed in 1952, Urey and Miller simulated the atmosphere of primitive Earth—a reducing gas mixture rich in methane and ammonia—and then added water and energy in the form of electric discharges. This led to the formation of amino acids and other organic substances. However, no experiment of this kind ever resulted in the formation of life. This led many biologists to wonder whether the primordial soup was actually sufficient and whether the scenario described above was reliable.

The environment was certainly very different from the probable real one, starting from the effect of water, which was not oxidizing as it contained very little oxygen, but reducing, and moving on to the atmosphere which, being reducing, probably contained much less methane and ammonia than that of Miller's experiments, but was essentially made of carbon dioxide and nitrogen. Volcanism was stronger than now and enriched the atmosphere in substances that are harmful to life today, such as sulfuric acid. Also, without an ozone layer in the atmosphere, ultraviolet radiation was very strong.

Between the ideas of life originating on the surface and in the depths of Earth, there is an intermediate hypothesis: life started on the surface, then migrated deep underground, and differentiated with the appearance of autotroph organisms that drew energy from Earth and not from the Sun. When a big asteroid hit Earth, causing the oceans to evaporate to a depth of several hundred meters, all living beings, except those protected deep under the seas or underground, were destroyed and life slowly had a new start, beginning from these survivors. As previously stated, even if life began earlier than 3.8 billion years ago, we will likely never know which of these hypotheses is correct; meteoric impacts canceled all traces of living beings that may have existed before Earth started to be a somewhat quieter place.

Astronomer Guillermo Gonzales once noted that theories on the origin of life endorsed by various scientists seem to depend on their specialization: oceanographers like hydrothermal vents on ocean floors, biochemists prefer tidal pools on the surface, astronomers underline the role plaid by comets in carrying complex organic molecules to Earth, and finally scientists who also write science fiction imagine that Earth was seeded by interstellar bacteria.[11]

In conclusion, in whatever way life started, it is certain that earlier than between 2.1 and 3.5, possibly even 3.8 or 4 billion years ago, extremely simple living beings, similar to current archeobacteria, existed on Earth. They were unicellular organisms, in which the genetic material, based on DNA, was situated directly in the cytoplasm, without the presence of a nucleus in the cell.

These beings remained the only inhabitants of our world for some billions of years.

---

[11] G. Gonzales, "Extraterrestrials, a Modern View," quoted in P.D. Ward and D. Brownlee, *Rare Earth*, Copernicus, New York, 2000.

PANSPERMIA

The difficulties encountered in explaining the origin of life on Earth led to the formulation of another hypothesis, according to which life arrived on our planet from space. Various versions of this hypothesis, generally known as *panspermia*, exist. In one case microorganisms arrived on Earth from the depths of space; in another, meteorites, detached from other planets by asteroid impacts, caused an exchange of living matter between the planets of the Solar System. Some scientists even support the hypothesis that life was purposely spread in the Universe by intelligent beings.

The first to speak of panspermia in modern times (the Greek philosopher Anaxagoras is said to have suggested a similar idea in antiquity) was probably Lord Kelvin; in a lecture in Edinburgh in 1871 he suggested that life had been carried to Earth by meteorites from planets on which it had developed in the past. The meteorites were assumed to have been launched into space by a violent shock. Today we know not only that meteorites can originate in this way, but also that it is possible that their inner parts survive the trauma of the formation process without serious damage.

This idea aroused incredulity, and it was immediately objected that any living organism would be destroyed by the heat due to air resistance during entry into Earth's atmosphere. The same objection was put forward by opponents of panspermia for more than a century, but, as already stated, we now know that it has a weak grounding. Despite these objections, the hypothesis had many supporters, such as the German physicist Hermann von Helmholtz.

This type of panspermia is often called litopanspermia; it gained support again after the discovery in 1996, in a meteorite from Mars, of structures that some identified as microfossils. At the start of the twentieth century Svante Arrhenius, the Swedish Nobel laureate for chemistry who probably introduced the term *panspermia*, suggested an alternative hypothesis: that microorganisms can be pushed through space by the pressure of solar light and travel not only inside the Solar System but also in interstellar space. Microorganisms could enter Earth's atmosphere without suffering from the extreme heat, but would they be able to withstand the space environment, hard vacuum, and strong radiation, for the lengthy periods needed for interplanetary or even interstellar journeys? In the latter case the journey could last many millions of years.

Some recent discoveries scored several points in favor of panspermia. Bacteria (*Micrococcus radiodurans*) able to survive very strong radiation—

10,000 times the dose a human being can withstand—have been discovered inside nuclear plants. The same bacteria have also been found in other places and, therefore, are not due to mutations induced by the artificial environment. Moreover, a thin layer of interstellar dust is able to protect the bacteria from the radiations of space. Old bacterial spores, hundreds of millions of years old, have been brought to life again and it seems that even a vacuum doesn't irreversibly damage the spores.

Proof of the ability of bacteria to survive in space was obtained when the astronauts of *Apollo 12*, Pete Conrad and Alan Bean, recovered and brought back to Earth the television camera of the *Surveyor 3* probe that had landed on the Moon almost three years earlier. The analysis of the camera showed that a colony of *Streptococcus mitis*, which evidently survived the sterilization procedures that always precede a launch, were still alive; the bacteria had withstood the vacuum, radiation, and strong temperature excursions on the lunar surface for 31 months.

Arrhenius's hypothesis was revived in the beginning of the 1970s by Fred Hoyle and Chandra Wickramasinghe, who affirmed that, thanks to accurate spectroscopical observations, they had obtained sure evidence that most interstellar dust is constituted by spores. Besides, they asserted again that comets carry life not only through the Solar System, but also in interstellar and intergalactic space.

Panspermia, at least in its initial form, obviously does not answer the question of the origin of life, it just moves the place of origin of life from Earth to an unknown planet in an unknown galaxy, as well as moving the event back in time by billions of years. But if it says nothing about the origin of life, it has a strong impact on its diffusion in the Universe and on the type of life that can be met on the various planets. If the panspermia hypothesis were true, life, at least at the bacterial level, would be extremely widespread and much more uniform than in the case of an independent origin on the various celestial bodies.

But the version of panspermia supported by Hoyle and Wickramasinghe also has a wider meaning: life didn't have any kind of *origin*, because it has always existed, like matter and energy (as mentioned earlier, Hoyle was one of the few scientists who do not accept the *Big Bang* theory). In the infinity of time that preceded us, life diffused through the whole Universe, giving origin also to an infinity of intelligent species. These species had all the time needed to improve the process of panspermia, designing microorganisms suitable to space conditions and purposely directing them toward the most suitable systems.

The two scientists went further. They, together with other supporters of this strong form of panspermia, stated that the mechanisms normally considered as the basis of evolution, that is, gradual mutations of the genoma of the various species, is wrong, or at least insufficient to explain the origin of new species. According to them, evolution proceeds only thanks to the introduction of new genes contained in viruses periodically entering the atmosphere from space. The viruses spread in the atmosphere and enter living cells, inserting new genes into them. By doing so, they allow evolution to go on. Therefore, if panspermia is directed by the will of the intelligent beings that preceded us, evolution develops according to a plan, through the insertion of purposely designed DNA chains. Apart from promoting evolution, viruses also naturally cause illnesses: Hoyle performed epidemiologic studies seeking to show that illnesses do not propagate owing to contagion, but that epidemics are linked to the entry of comets into the Earth's atmosphere.

Further generalizations of the panspermia hypothesis include James Lovelock's *Gaia hypothesis* (see page 103); this widened version is also known as the *cosmic ancestry* hypothesis. If science is open to weak versions of panspermia and does not rule out the possibility of the propagation of life from one celestial body to another, it is, on the other hand, extremely skeptical of the strong versions. It should at least be noted that such hypotheses would require really strong evidence, which nobody has so far succeeded in producing.

In support of panspermia, recent discoveries of fossils in meteorites of the carbonaceous condrites type have been reported. As already mentioned, structures similar to microfossils were identified in the Murchison meteorite. In July 1999, Stanislavs Zhmur and Lyudmila Gerasimenko, of two institutes of the Russian Academy of Sciences, announced the discovery of structures similar to microorganisms (cyanobacteria of the *Lyngbya oscillatoria* species), with traces of cellular structure, in the same meteorite and in the Efremovka meteorite (Figure 3.7). The same authors stated that spherical particles found in the specimens carried back to Earth by the automatic probe *Luna 20* in February 1972 look like bacteria of the *Siderococcus* or *Sulfolobus* type, and that fossils were also found in the specimens carried back to Earth from the Moon by *Luna 16*.

Claims of discoveries of fossils in material from space are being made at a growing pace. Most scientists are skeptical and hold that the identification of the structures is not quite certain and that it is not possible to rule out contamination by terrestrial microorganisms, objections similar to those that have been raised against the discovery of microorganisms of Martian origin in a meteorite of the LSC type, announced in 1996. This claim and the relative objections will be discussed later, when dealing with the possibility of life on Mars.

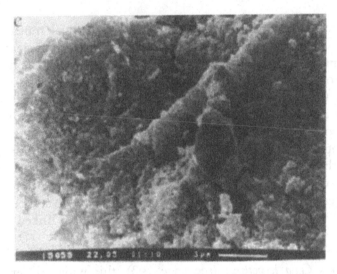

FIGURE 3.7 Filamentous fossil microorganisms, with traces of cellular structure, similar to cyanobacteria, allegedly found in the Murchison meteorite. (From the Web site www.panspermia.org/zhmur1.htm.)

In conclusion, it can be stated that none of the discoveries of material of biological origin in carbonaceous condrites or in space was confirmed beyond doubt, and that panspermia, at least in the form of litopanspermia, is a sound hypothesis, but still needs convincing evidence.

EVOLUTION AND CREATIONISM

In the preceding sections (apart from that dealing with panspermia) the history of the beginning of life on our planet was reconstructed on the basis of what is today considered one of the most reliable paradigms of science, the theory of evolution introduced by Darwin and, before him, by other nineteenth-century scientists. This formulation is endorsed by almost all scientists, but a nonnegligible part of public opinion still opposes it, sometimes violently, in the name of a creationist interpretation of the origins of life. Is must be expressly stated that creationism is opposed to the very concept of evolution and not just to the specific form of the theory known as Darwinism—or better, to its present form, Neo-Darwinism.

The idea that animal and plant species transform, adapting themselves to the environment, was actually not introduced by Charles Darwin. He and Alfred Wallace, who simultaneously and independently came to the same conclusions, improved and put on a more sound basis some ideas that Buffon, Lamarck, and other philosophers and naturalists of the eighteenth and early nineteenth centuries had already put forward. In particular, Darwin accumulated a great deal of observational evidence that strengthened his hypotheses. Darwinism is based on the idea that evolution occurs gradually, through small random variations in living things, and that natural selection performs the task of choosing, among the many changes that have taken place, those that, because they represent an advantage from the point of view of survival, will remain as a legacy for the following generations.

In this way, all traces of finalism are eliminated: variations happen at random and only later is a choice made between those introducing an advantage and those, much more abundant, that are instead harmful to survival. The term *natural selection* was introduced by Darwin under the influence of the process of artificial selection that farmers use for improving vegetable and animals species, with the difference that, in that case, the procedure is driven by a precisely determined will, while in nature it is driven by chance.

The proper understanding of the term *evolution* does not entail implication of "progress" or of "betterment": a new species is simply different from that from which it derives, but it is neither better nor more evolved. Surely it is more suited to the environment, but this cannot constitute an absolute term of comparison, since the environment continually changes and therefore what is more suited today may not be so tomorrow. There is a difficulty in this formulation, since the study of fossils shows that evolution always proceeds toward a greater complexity; however, complexity may be interpreted not as a symptom of progress but simply as a result of the number of attempts continuously increasing in time.

Darwin had little knowledge of genetics and therefore could not deal with the details of the mechanisms producing the changes whose effects he observed. Genetics, developed mainly in the second half of the nineteenth century by Mendel, was meanwhile proceeding in a completely independent way. Neo-Darwinism, which developed in the twentieth century, resulted from the introduction of genetics into Darwinism and allowed the organization of our knowledge of the evolution of life on Earth into a consistent theory. But if Neo-Darwinism can explain microevolution, that is, the changes that cause small variations, adjustments to environmental conditions for which the theory has been experimentally validated, the detailed mechanism of macroevolution and, as stated, of the origin of life, are still unclear.

The evolutionist interpretation has been extended from the living world to the totality of the Universe and now, in addition to biological evolution, we speak of cosmic evolution, stellar evolution, chemical evolution, etc. It must be remembered that since its first formulation by Lamarck, the theory of evolution aroused strong opposition, above all from those who felt outraged by the fact that humans lost their specific, privileged status of living, being completely separate from the animal world. After all, the most strongly contested statement of the theory of evolution is the one summed up in the following sentence: *Man descends from apes.* Christian churches and many exponents of other religions, but also many scientists, immediately rallied against the theory of evolution.

Religious opposition to the theory of evolution decreased with time, to the point that Christian scientists themselves developed evolutionary hypotheses, generally characterized by an underlying finalism: the term *evolution* again takes on a meaning of progress, of tendency toward a goal. But this obviously goes beyond science, since considerations on the possible finalities of the process under study do not belong to the scientific sphere. The Catholic Church expressed its position in an uncompromising way with a sentence pronounced in 1996 by Pope John Paul II: "New discoveries lead to recognize that the theory of the evolution is more than just a hypothesis." Other Christians churches, on the contrary, are still rigidly entrenched in creationist positions. In general, the creationist interpretation of life and the Universe states that all that exists, including the various forms of life populating the Earth, was created by God exactly as it is at present, in a single creative act.

The fact that some species became extinct does not pose a problems to this interpretation; what is instead problematic to explain is the age of Earth in comparison with that of humanity. If all living species were created at the same time, then men, dinosaurs, trilobites, and every other living being must have all lived together, at the same time, on this planet. Since it seems absolutely unrealistic that humans have been present on Earth for almost four billion years, creationists were forced to challenge the whole geological dating of Earth and to state that our planet is much younger than scientists believe. Moreover, creationism generally supports a literal interpretation of the Bible and therefore the dating deriving from it, too. The creation of the world and of all living beings is assumed to have occurred some thousands of years before Christ, at most about 10,000 years.

Creationism is therefore in contradiction with the whole of modern science, not only with biology. If Earth is only a few thousand years old, all of geology and astrophysics and a good part of physics and chemistry are drastically wrong. Sometimes one wonders how it is possible to advocate positions of this kind. Actually this is not the first time in history that scientific theories, generally accepted and supported by experimental evidence, are challenged in the name of truths not based on alternative scientific hypotheses but on beliefs extraneous to science.

This had not been the case in the debate between the geocentric and the heliocentric theories in astronomy: in that case two theories existed (or better, three, since an intermediate theory, conceived by Tyco Brahe, was also present), and in the beginning the heliocentric one was less in agreement with astronomical observations. But even later, after Kepler introduced his three laws, Galileo demonstrated with his experiments that the rotation of Earth was in agreement with the laws of physics and the evidence in favor of the heliocentric theory became overwhelming, many in the Catholic and Protestant churches continued to support the geocentric theory, basing this on a literal interpretation of the Holy Scriptures.

Another interesting intellectual episode tied to the subject of this book is the theory of the spontaneous generation of living creatures. The fact that not only bugs and other small animals, but also mice and other mammals, can spontaneously take life from inorganic matter was commonly accepted in the ancient world. The idea survived all scientific revolutions and theological debates until a little less than two centuries ago. Thinkers like St. Augustine and St. Thomas, writers like Shakespeare and Goethe, and scientists like Galileo and Copernicus were convinced of it and did not see in this theory any contradiction with contemporary science or religion. As late as the seventeenth century, biologists held that if one put cloths dirty with sweat and wheat in an open vessel, the wheat would slowly transform into adult mice, of both sexes, and that such mice, indistinguishable from those born in the usual way, could couple with other mice and generate new animals similar to their parents.

It slowly become clear that the spontaneous generation of complex animals was impossible, yet in the nineteenth century many biologists still believed in the spontaneous generation of microorganisms. By halfway through the nineteenth century it had been discovered that living cells were mainly constituted by protoplasm, and since protoplasm was considered to be a simple substance that was potentially obtainable from nonliving matter, this strengthened the idea that the simplest micro-organisms could originate spontaneously.

Performing—and, above all, interpreting—experiments that could settle the matter was not at all simple. An example is the quarrel, in the second half of the eighteenth century between the Welsh Needham and the Italian Spallanzani, both scientists and both Catholic priests (the first was a Jesuit). Needham carried out various experiments in which he sterilized vegetable infusions by sealing and boiling them, and observed the formation of microorganisms (animals of the infusions, today called infusors). Spallanzani repeated the experiments, but boiling for a longer time, and did not observe any form of life—clearly evidence against spontaneous generation. To this Needham replied that the treatment used by Spallanzani had weakened, if not completely destroyed, the *vegetative force* of the substances in the infusion. In other words, a different interpretation of the experiment led to opposite conclusions.

Finally, in the second half of the nineteenth century, Pasteur's experiments clearly showed that spontaneous generation of microorganisms did not exist, but the quarrel continued until the end of the century. At that point it was no longer an academic debate: to advocate the theory of spontaneous generation was also to deny the usefulness of sterilization in surgery or of vaccination in medicine. The fact is that since the beginning of the nineteenth century the quarrel on spontaneous generation had assumed a character that went well beyond that of a scientific debate. To use the words of Pasteur in a lecture at the Sorbonne:

> What a victory for materialism if it could be affirmed that it rests on the established fact that matter organizes itself, matter which already has in it all known forces! ... Of what good could it be then to have recourse to the idea of a primordial creation, a mystery before which it is necessary to bow? Of what good would the very idea of a Creator God be?

For many, after the French Revolution, the theory of spontaneous generation meant scientific materialism, atheism, the Republic, revolution, terror, the guillotine. The other side proclaimed that "denying spontaneous generation amounts to proclaiming a miracle." And speaking of miracles meant denying freedom of thought, sustaining a church often heavily involved in politics, supporting the aristocracy, social oppression, the persecution of those whose ideas were different from what the establishment endorsed, and so on down to the crusade against the Albigenses and the Inquisition.

When a scientific debate degenerates in this way, the most elementary conditions for free and impartial research vanish and fanatics feel that any action is justified whose goal is to make their ideas triumph: after all, if the stakes of the game are the future of society—they think—creating some evidence and adjusting the results of experiments may well be excused!

Today the arguments against spontaneous generation are again employed by supporters of the strong form of panspermia, particularly some statements by Pasteur on the possibility that life had always existed, together with matter. But to affirm that life evolved slowly from non-living matter is clearly quite different from supporting spontaneous generation, and the evidence presented against that theory does not carry weight in the discussion on the origin of life on our planet.

Once the theory of evolution had been formulated, the debate on spontaneous generation went on, in the form of debate between evolutionism and creationism. In this case the emotional impact was much greater, because the theory involves humans. This is the reason, as mentioned earlier, that many people find it more difficult to accept evolution: as long as it limits itself to assessing that an animal species derives from another species, there are no great difficulties, but that human beings should be linked to the animal world by direct descent is a rather more difficult concept to accept.

Even today, after the emotional impact has dampened and the evolutionist view has been, so to speak, metabolized, the polemic between creationism and evolutionism continues, though in an increasingly marginal way. The creationists, paradoxically, accuse science of being prejudiced and supporting a materialistic and atheistic worldview at all costs. They accuse scientists of improper scientific behavior and of ignoring, or destroying, evidence that goes against the official truth. It is ironic that the words used by creationists to accuse science should echo those used by free thinkers against the Church a century ago.

In a creationist view of the world, the existence of extraterrestrial life or intelligence is not a scientific problem but a theological one: some creationists actually do not deny that life can exist in other places in the Universe, but for its existence a deliberate and specific creative action of God is required.

TOWARD A GREATER COMPLEXITY

As already mentioned, the fact that life appeared on our planet in such a short time and the extreme variety of environments that can sustain it lead us to think that it is not at all an unlikely occurrence in the Universe, at least as far as the simplest living beings are concerned. But after the first prokaryotes appeared on our planet a long period of stasis ensued, more than a billion years in which nothing new seems to have happened. The prokaryotes never formed true multicellular organisms, even if colonies of unicellular beings were present since the beginning. Stromatolites as old as

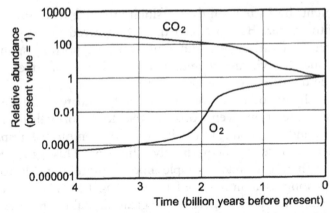

FIGURE 3.8 Concentrations of oxygen and carbon dioxide in Earth's atmosphere as a function of time. The values are given in comparison to the present situation. Notice the logarithmic scales: at the beginning the concentration of the oxygen was, in comparison to the present, less than one ten thousandth, while carbonic dioxide was a thousand times more abundant. The values for the most ancient times are only estimates, but the trend is accurate.

3.5 (or 2.1, according to other authors) billion years are among the most ancient fossils; these are fossilized structures formed when thick layers of blue algae growing in shallow waters were covered by inorganic deposits, then by new layers of algae, and so on. Other fossils of the same age are simply microfossils—that is, very small marks left by unicellular organisms.

Note that it is very difficult to recognize fossils, especially microfossils, from sometimes very similar structures, created by phenomena that have nothing to do with living matter. This is true for the supposed microfossils found in meteorites, but also for similar discoveries in terrestrial rocks.

Fossils that are assumed to be more recent are more abundant and their interpretation less uncertain. It is, however, taken for certain that life made very little progress for one or two billion years, always reproducing the same scheme, that is, bacteria or other unicellular organisms based on the structure of the prokaryotes. In the meantime those simple and primitive beings were performing a cyclonic task whose result was a radical change of the planet: they replaced an atmosphere made mainly of carbon dioxide and nitrogen with one based on oxygen and nitrogen (Figure 3.8).

The largest part of the job was performed about two billion years ago, mainly by unicellular algae, and lasted for several hundred million years, but the carbon dioxide content continued to decrease, almost up to the present. Even in relatively recent times (geologically speaking), such as the period in which dinosaurs lived (from 200 to 70 million years ago), the atmosphere contained much more carbon dioxide than it does today.

The change in the composition of the atmosphere was without doubt a traumatic event and caused the extinction of many species for which oxygen was actually a poison. The only organisms that survived were those living deep underground or in the depths of the sea, where the quantity of oxygen dissolved in water was sufficiently small. Other species adapted to life on a planet rich in oxygen, and new species that used oxygen directly began to evolve.

Between 2.5 and 1.7 billion years ago (here too it is difficult to assess precisely when, and different estimates exist) another change occurred that would have important consequences for the evolution of life: the first cells appeared in which the genetic material was included in a nucleus, separated from the rest of the cellular material by a special membrane. These cells also had other internal differentiations (mitochondria, plastids, etc.). The inner structure of these living beings, collectively called eukaryotes, is much more complex than that of the prokaryotes and their evolutionary potential was clearly much greater.

It seems that eukaryotes originated as a sort of symbiosis between cells. Prokaryote bacteria entered into another bacterium and later transformed into the mitochondria, plastids, and perhaps even into the nucleus, which in the beginning would have been a kind of cell within a cell. Subsequently, part of the genetic material of these inner cells would have been lost while part was transferred to the nucleus, in which the whole genetic material of the cell finally gathered. With eukaryote cells, sexual reproduction started and the path that would lead to multicellular organisms was opened.

Meanwhile, Earth's environment was undergoing important transformations. The Sun, like all stars, increased in brightness as it grew older and stronger, causing frequent climatic changes. It is likely that at the beginning the temperature of Earth was much higher than at present, owing to the high carbon dioxide content causing a strong greenhouse effect, but some scientists hold that it was, on the contrary, lower due to lower solar heating. For long periods, the first one beginning about 2.45 billion years ago and then again between 800 and 500 million years ago, the temperature decreased so much that the planet was transformed into a ball of ice. The glaciations that occurred in more recent times were marginal episodes in comparison and, above all, less global. It seems that, at least in the first of such periods, the sea was completely frozen to a depth of some hundred meters and the land, if some land was present, was covered by ice.

The composition of the atmosphere was undergoing radical changes, thanks in part to living beings but also to other phenomena—like, for example, volcanism—that periodically introduced millions of tons of carbon dioxide, sulfuric acid, and other gases into the atmosphere. Perhaps it was the greenhouse effect, due to gases coming from the volcanoes, that put an end to the various periods of intense cold. When this happened, the melting of the ice caused large quantities of iron and magnesium contained in the dust covering the ice to enter into the sea. These metals acted as a fertilizer, causing a large development of blue algae and fitoplankton. This in turn caused large quantities of oxygen to be liberated in the atmosphere: the whole process may have had a powerful effect on the development of many living species.

The beginning and end of the cold periods, and the subsequent rapid variations of the environment, may have caused the first mass extinctions and the development of new species, thus accelerating evolution. In this case the evolution of life would have been helped more than hindered by the sudden climatic changes.

It is not known exactly when the first eukaryote cells appeared, but the event seems to have had little consequence for several hundred million years. In fact, even the appearance of the first multicellular organisms (800 million years ago?)—in which cells started to specialize, assuming diversified functions—did not cause an immediate revolution.

Until half a billion years ago there was no sign of life on dry land, and even in the sea the only living beings were microorganisms and some simple animals, probably similar to worms. Apart from the color that large masses of microorganisms may have given the water, Earth would have looked like a dead planet, like the Moon or Mars today. Then, about 550 million years ago, evolution underwent an abrupt acceleration, with the appearance of animals and plants in rapid sequence. In a relatively short time, almost all types of living beings that populate the Earth today appeared: from jellyfish to fish, from arthropods to Chordata, a phylum[12] to which we also belong. Living beings of large dimensions, at least in comparison with the microscopic organisms that existed earlier, appeared for the first time.

This unbelievable expansion of life on our planet is generally called the *Cambrian explosion*. It left many fossil-rich strata in the whole planet, in stark contrast with the preceding strata, in which evidence of life is so scarce that, until a few years ago, life was thought to have started at the beginning of the Cambrian. Today we know that it is not so, and that life

---

[12] The phylum is a very large subdivision—for instance, all vertebrates, from the fish and the reptiles to humans, belong to the phylum of the Chordata, which also includes three other subphyla.

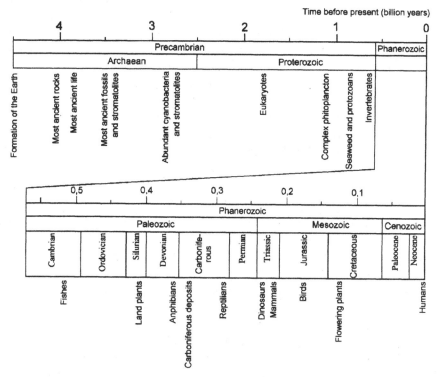

FIGURE 3.9 *Synthetic chronology of life on Earth, deduced from geological evidence. The eras and, for the Phanerozoic era, the periods, are shown (the figure follows the hypothesis that life originated about 3.7 billion years ago).*

is far more ancient, but there is no doubt that life as we know it began with that geological period (Figure 3.9).

The present number of animal species is estimated between 6 and 30 million, but all this variety can be grouped in just 28 to 35 fundamental types (phyla). Almost all these types originated in the Cambrian era, together with many other phyla that later became extinct. Although it is not possible to evaluate their exact number, it seems that at the end of that period there were about 100 phyla. The Cambrian almost seems to have been a time of intense experimentation, in which nature created an enormous number of possible animal and plant configurations. Subsequently evolution experimented with endless variations of these themes and selected the configurations that were most suited to the various environments. Therefore, while the number of species continued to increase, the number of phyla decreased.

From that time the transformation of the planet was quick and radical: about half a billion years ago life emerged from the water and started to colonize the dry land. Four hundred million years ago, in the Silurian era, the first arborescent vegetables appeared. The face of the continents had remained almost the same for a billion years, although they had changed in form and extension, but now in a few million years they were covered by vegetation and populated by animal life of every kind. In little more than 150 million years, the dry land saw an impressive development of animal life: first it was just fish timidly crawling out of the water but quickly larger animals developed, up to the gigantic dinosaurs. Just 50 million years later reptiles were learning to fly, soon to evolve into birds.

CATASTROPHES AND MASS EXTINCTIONS

The development of life must not be thought of as a linear process, a number of small steps performed one after the other, even if not at a constant pace, in a progression (or, if one does not accept any finalism underlying evolution, in a change) leading from inorganic matter to human beings.

Geological changes are extremely slow. Those occurring today—a typical example is the continental drift—could not be perceived by us if we were not using sophisticated instruments. However, evidence has been found of quick and violent phenomena that caused the extinction of both innumerable individuals and whole species, and the creation of new evolutionary opportunities for the surviving species. The percentage of marine species that became extinct in various episodes of mass extinction, beginning from the Cambrian, is shown in Figure 3.10.

Such phenomena can be of various types and may originate from inside the planet or from space; their study is just beginning and little is known about them today, except for the effects they caused: mass extinctions in which up to over 90 percent of species—and probably an even higher percentage of individuals—disappeared.

The mass extinction best known to the public is the one experts call the *K/T extinction*, which at the end of the Cretaceous period, about 65 million years ago, destroyed 85 percent of animal species and put an end to the age of the dinosaurs. Its consequences were probably even greater than just the end of the huge reptiles; without it, the evolutionary niches that were later occupied by mammals would probably not have been left free, and it would hardly have been possible, millions of years later, for higher primates and then humans to develop at all. Without the K/T extinction, intelligence would probably not have developed on our planet, or it might have developed in a different class of being, possibly a reptile. Still, biology

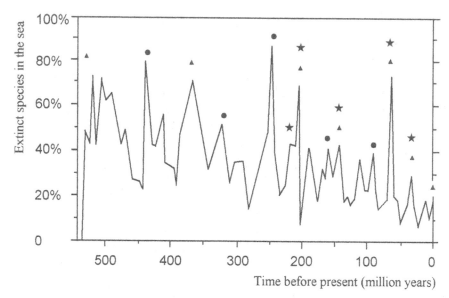

FIGURE 3.10 *Percentage of marine species that became extinct in various episodes of mass extinction.* ★ *large impact crater,* ▲ *clues of impacts,* ● *iridium layers of lesser importance.*

FIGURE 3.11 *The impact of the asteroid that caused the K/T event in which the dinosaurs, together with a large number of species, were extinguished. (NASA image.)*

87

can no more be built on *ifs* than history, and the K/T extinction happened, with all its consequences.

It is now widely accepted that the K/T extinction was caused by the impact of a big asteroid or comet, an object with a diameter of about 10 km that struck Earth near Chicxulub, in the Yucatan (Figure 3.11).

The first clues that the K/T event was caused by an impact were discovered near Gubbio, in Italy, where paleontologists discovered a thin layer very rich in iridium, formed about 65 million years ago. Iridium is a very rare metal on the Earth, but quite common in meteorites, so the idea that a large asteroid had struck our planet was advanced. The fact that dinosaurs disappeared at the same time suggested that the two things were correlated.

At the beginning this hypothesis met with much disbelief. Paleontologists always tend to resort to slow, long-lasting processes and are suspicious, with reason, of hypotheses invoking catastrophes, particularly when the probability is very low. Moreover, there is a tendency to forget that Earth belongs to a wider Universe and that what happens "out there," in the Solar System and beyond, may have heavy consequences on the events occurring "at home." The extinction of the dinosaurs was assumed to be a gradual phenomenon, possibly lasting hundreds of thousand years, caused by events that originated on Earth itself.

Yet slowly the clues of a meteoritic impact accumulated and finally the decisive evidence was discovered: an impact crater with a diameter of about 200 km, much damaged and partly submerged under the gulf of Mexico today, but dating from the correct time. That the K/T extinction, which occurred between the Cretaceous and the Paleocene (Tertiary), was caused by an asteroid is therefore much more than a simple hypothesis now.

When a celestial body of that size strikes a planet, a huge quantity of energy is released, of the order of 10^8 (one hundred million) megatons (a megaton is the energy developed by a million tons of TNT). By comparison, the atomic bomb that destroyed Hiroshima had a power of a hundredth of one megaton and the power of the largest hydrogen bombs is in the range of 50 megatons.

The heat developed in the impact is so intense that if the asteroid falls in the sea the surface layers of water instantly evaporate within a range of hundreds of kilometers around the point of impact. If the impact occurs on land, it causes the projection of an enormous quantity of incandescent fragments. The resulting crater has a diameter about 20 times that of the asteroid; no form of life can survive in the area where the crater is formed and its immediate neighborhood. Some of the fragments, those possessing a high enough speed, are projected into space, while most of them return to Earth, even at very large distances. Almost the entire planet is affected by the fall of glowing fragments, which set fire to most of the vegetation.

The large quantity of dust projected into the atmosphere, together with the smoke from the fires and the products of the likely volcanic eruptions that would be triggered by the impact in surrounding zones, form a thick layer in the high atmosphere that stops solar radiation, causing the temperature to drop. The effect is something similar to the so-called nuclear winter that could follow the explosion of many nuclear weapons, but on a much greater scale. Darkness and the nuclear winter, in these conditions, may last over a year, causing animals that survived the destruction of the impact and the following fires to starve. The vegetation that survived the fires and the ensuing cold temperatures is damaged by the acid rains caused by nitrogen and sulfur oxides, which produce nitric and sulfuric acid. When the layer of dust finally deposits and sunlight is able to filter through the atmosphere again, things may get even worse, since the ozone layer is at least partially destroyed and the surface is exposed to strong ultraviolet radiation. Water, acidified by the acid rains, is poor in organic substances and so allows a greater penetration of ultraviolet radiation into the sea environment.

The asteroid responsible for the K/T event struck a zone rich in sulfur compounds. This, on the one hand, made things worse, increasing the acid rains, but, on the other hand, produced a haze lasting for years, which reduced the damage caused by ultraviolet radiation. Perhaps if the asteroid had struck another zone, less rich in sulfur, things would have been even worse. An asteroid of a larger size would have caused a severe loss of atmosphere and a strong evaporation of the oceans, whose water would be mostly lost in space (Figure 3.12). If this had happened, life could have been completely erased from the planet.

Events of this sort can easily occur in the early phases of a planet's life, but the arrival of comets at a later stage can bring new water and volatile substances, forming new seas and a new atmosphere. If they happen later, however, the planet may remain barren forever. There are clues, of unknown reliability, that something of this kind happened to Mars. Actually it is now almost certain that Mars originally had a dense atmosphere and a lot of water. It is possible that a large asteroid struck the planet (the crater *Hellas* or *Vastitas Borealis*, a large depression occupying almost all the northern hemisphere, could be the result of the impact), causing the seas to evaporate, destroying the atmosphere, and reducing the planet to the conditions we observe today.

Other asteroid impacts were responsible for other mass extinctions, starting from those that, as already stated, may have forced life to a new beginning—or at least to hide underground or in the depths of the oceans to survive. Beginning with the Cambrian, a periodicity of mass extinctions can be noticed, with similar events occurring every 30 million years or so. In many cases impact craters or iridium layers exist that can be dated, at least approximately, to the same time frame of the extinction (Figure 3.10).

The observation of this periodicity started the so-called *Shiva* hypothesis: owing to the periodic crises caused by the fall of asteroids or comets on Earth, life is periodically almost canceled and then starts again with renewed strength. It is not known why these celestial bodies should fall periodically on Earth, but various hypotheses have been advanced. In particular, it has been suggested that the Sun might be a double star, with a very dark (perhaps a brown dwarf), and hence not yet observed, small companion, with a very elliptical orbit. Periodically, Nemesis, as this star has been called, approaches the Sun and disturbs the Oort cloud, diverting a large number of comets toward the inner Solar System, on orbits crossing those of the inner planets and therefore also of Earth.

Another hypothesis postulates an oscillation, with a period of about 26 million years, in the motion of the Sun around the center of the galaxy—a

FIGURE 3.12 *Impact of a large asteroid on Earth. In the first instant a large fraction of the atmosphere is projected into space. (NASA drawing.)*

revolution with a period of about 226 million years. This oscillation causes the Solar System to move above and below the mid-plane of the galaxy, periodically crossing a zone more densely populated by stars that can disturb the Oort cloud, sending comets toward the inner Solar System. At present, this is just a hypothesis, needing further study.

Other mass extinctions that could be ascribed to the impact of an asteroid or a comet are those that occurred around 440, 250, and 202 million years ago. In particular, the extinction located between the Permian and the Triassic periods, 250 million years ago, was the most catastrophic of all, making perhaps more than 90 percent of species extinct. As an alternative to the fall of an asteroid, it has been suggested that this was caused by a strong increase of volcanic activity, with large quantities of sulfur and carbon dioxide erupted by volcanoes directly into the atmosphere, together with more carbon dioxide that had previously been dissolved in the oceans. In this scenario the strong greenhouse effect and consequent increase in temperature, accompanied by acid rains, would have caused the extinctions.

The impact of even a small asteroid may have severe consequences on life on our planet: which means that we would be well advised to evaluate the danger to which we are currently exposed from such an event. First, it must be stated that while the danger today is very small compared to that at the beginning of Earth's history, when so many small bodies that had not yet joined a planet were roaming free in the Solar System, the fact is that the situation has not substantially changed in the last 65 million years: on an astronomical time scale, the danger we face today is no less than that faced by the dinosaurs. The probability that an asteroid like the one that caused the K/T extinction should fall on Earth within geologically long times is almost equal to 1, that is, the event is certain to occur. Our saving grace is that the probability of it happening within our life span is very small.

The average time interval between the fall of two asteroids of the same type is presented in Figure 3.13 as a function of the diameter of the object (or the energy of the impact). One can see that an asteroid the size of the one that fell 65 million years ago is encountered by Earth less than once every 100 million years, but asteroids still capable of having global effects are much more frequent, less than one in a million years.

Yet humanity is very vulnerable to events of this type, and even to lesser ones that, although not causing true mass extinctions, would in any case cause very serious damage. The fall of a meteorite or a small cometary nucleus like the one that fell in the beginning of the twentieth century at Tunguska in Siberia has a probability of about one event each century and

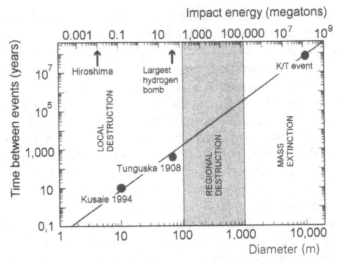

FIGURE 3.13 *Relationship between diameter of the asteroid (or the energy of impact) and the mean time between two events.*

does not jeopardize any living species. Incidentally, the fact that it happened almost a century ago gives us—even those of us who know that statistics cannot be interpreted in this way—something to think about. Yet if the Tunguska asteroid had struck a densely populated area instead of a desert forest, it would have caused the death of thousands or even millions of people and huge economic damage. Again, statistics are in our favor: the probability that a city is struck is much lower than that of the asteroid falling on a sparsely populated area and, in the absence of global consequences, this is somewhat reassuring.

An event of this type has no biological consequences, and from this point of view the life of the individual does not matter; what counts is only the survival of the species. However, since the time an intelligent species appeared on Earth, the survival of individuals became a primary goal and localized catastrophes must be considered a danger to be avoided.

It is clear then that if an intelligent species wants to survive for a long time, it must learn to face events of this type or avoid them, for instance, by diverting or destroying asteroids on a collision course with its planet, or to minimize the consequences of the impact by selecting other celestial bodies on which a large enough number of individuals could survive. But to protect the life of individuals and not only the survival of the species, it is necessary to prevent even the fall of small asteroids.

It should be noted that the probability of being involved in an asteroid collision is, for each of us, extremely low—but not much lower than the probability of being involved in accidents of other types, such as an aircraft crash; if aircraft accidents are much more frequent than the fall of asteroids or comets, the number of casualties they cause is smaller by orders of magnitude and these two factors tend to balance each other.

The seriousness of the impacts of meteorites and asteroids has been codified in a scale, called the Torino scale (Table 3.4). This scale, conceived by Richard P. Binzel, was presented to the United Nations in

TABLE 3.4. *The Torino scale*

White Zone: Events having no likely consequences

0   The likelihood of a collision is zero, or well below the chance that a random object of the same size will strike Earth within the next few decades. This designation also applies to any small object that, in the event of a collision, is unlikely to reach Earth's surface intact.

Green Zone: Events meriting careful monitoring

1   The chance of collision is extremely unlikely, about the same as a random object of the same size striking Earth within the next few decades.

Yellow Zone: Events meriting concern

2   A somewhat close, but not unusual encounter. Collision is very unlikely.

3   A close encounter, with 1 percent or greater chance of a collision capable of causing localized destruction.

4   A close encounter, with 1 percent or greater chance of a collision capable of causing regional devastation.

Orange Zone: Threatening events

5   A close encounter, with a significant threat of a collision capable of causing regional devastation.

6   A close encounter, with a significant threat of a collision capable of causing a global catastrophe.

7   A close encounter, with an extremely significant threat of a collision capable of causing a global catastrophe.

Red Zone: Certain collisions

8   A collision capable of causing localized destruction. Such events occur somewhere on Earth between once per 50 years and once per 1000 years.

9   A collision capable of causing regional devastation. Such events occur between once per 1000 years and once per 100,000 years.

10   A collision capable of causing a global climatic catastrophe. Such events occur once per 100,000 years, or less often.

1995 and officially accepted in its final version during a conference held in Turin in 1999. It is a scale of the type of the Richter scale for earthquakes, aiming to classify the danger represented by the asteroids and comets that are discovered. It has 11 degrees (from 0, no risk, to 10, certainty of generalized destruction), and five colors (from white, certainty of no impact, to red, certainty of impact).

The usefulness of this type of scale derives from the fact that when an asteroid is discovered, its orbit can be calculated only in an approximate way and the potential danger it represents may be assessed only in statistical terms. Even when the orbit is better known, it can be changed by gravitational perturbations of the planets or even of Earth in such a way that the danger it represents changes. These perturbations cannot be computed with the required precision: a variation of a few thousand kilometers, a trifle on an astronomic scale, may transform a harmless asteroid into a serious danger.

Up to June 2005, 3428 objects (3371 asteroids and 57 comets), whose orbits approach that of Earth (NEO, near Earth objects; NEA, near Earth asteroids; and NEC, near Earth comets), were known, 783 of which have a diameter equal to or larger than 1 km and are potentially able to cause a global catastrophe (see Figure 3.13).[13] Also, 701 asteroids have been classified as potentially dangerous (PHA, potentially hazardous asteroids), because they could come dangerously close to Earth. This does not mean that sooner or later they will fall on us, but only that they are potentially able to do so. None of the known objects has a degree higher than 1 on the Torino scale. An example of an asteroid with a Torino scale value of 1 is 2004 MN4 Apophis, which will make several close passes, coming very close in 2029, and has some probability of hitting Earth on April 13, 2036. Its estimated diameter is 390 m.

Some events that occur in space are far more catastrophic than the fall of an asteroid or a meteorite; they may have had serious consequences on life on Earth and may have completely canceled life on other planets. One such event is the explosion of a nearby nova or supernova. If the star around which a planet orbits explodes, everything in the system is wrecked, to the point that even the most distant planets are subjected to temperatures and radiation that would not only kill every living being but also wipe out every material structure. Nearby systems within a radius of tens of light-years would also be flooded by such strong radiation as to cause the complete extinction of every form of life, while at greater distances, within a radius of hundreds of light-years, serious mass extinctions would be caused. The

---

[13] http://neo.jpl.nasa.gov/neo.

seriousness of the effects depends on the distance, on the mass of the exploding star, on the type of radiation released, and on the ability of the planet to screen its surface by, for instance, an intense magnetic field. There is no evidence that any of the mass extinctions on Earth were due to this cause, and this is in accordance with the fact that no residues of exploded stars have been found close (astronomically speaking) to the Solar System.

A cosmic occurrence even more dreadful than the explosion of a nova is a gamma-ray event. Sudden emissions of gamma-rays have been recorded, luckily in very distant galaxies; they are extremely powerful, and they are able to eliminate life in an entire galaxy and even affect nearby galaxies. Events of this kind are very rare indeed, but their destructive potential is enormous.

Another type of dangerous event is the strong emission of cosmic- and gamma-rays that follows the collision of two neutron stars; when two objects of this type merge, enormous quantities of energy are liberated in a few seconds. If a beam of high-energy particles created in this way should flood the atmosphere of our planet, it would destroy the protective layer of ozone and expose the surface to such a high dose of radiation as to destroy most life on dry land. It is possible that some of the mass extinctions were caused by events of this type, but it is not easy to find any evidence; moreover, we do not know either their periodicity or their actual danger. The physicist James Annis holds that a collision of two neutron stars in the center of our galaxy could destroy life on Earth and on any other possibly inhabited planets of the Milky Way. Since Annis calculated that an event of this type occurs in each galaxy every few hundred million years, and that in the past they were even more frequent, this makes it difficult to explain how life managed to survive on Earth up to the present time. These cosmic- and gamma-ray emissions are probably much less dangerous, for instance, because (a) they cause very serious damages in one hemisphere only—the one directly exposed, or (b) they are much less frequent. If their frequency is indeed very high in young galaxies, complex life could be a relatively recent phenomenon.

However, mass extinctions are not only due to the impact of asteroids and meteorites or to other cosmic events; causes are also linked to the geologic evolution of our planet. As already mentioned, the extinction that occurred about 250 million years ago, for instance, could have been due to intense volcanic activity. The emission of large quantities of carbon dioxide by volcanoes can cause an increase in the greenhouse effect and a generalized increase in temperature, while large quantities of dusts and solid material in the upper high atmosphere may have the opposite effect of cooling the planet, in what has been called a nuclear winter. The two effects may also combine, since they have different time scales: a rapid cooling followed by a slower but intense heating.

The history of the animal species in the last tens of thousands of years suggests a further cause of mass extinction: the appearance or, better, the generalized spreading on the planet of an intelligent species. It should be noted, however, that contrary to what most people think, modern humans are not the main culprits of the present mass extinction, affecting thousands of animal and vegetable species. The phenomenon started between 40,000 and 20,000 years ago, when human cultures were very primitive. The absence of large animals in Australia, for instance, is a consequence of the extinction of the large marsupials that occurred about 15,000 years ago. The cause of this extinction is not certain, but the fact that the first humans started to populate that continent around that time is unlikely to be just a coincidence. A similar phenomenon occurred in America, where many species disappeared after the arrival of the first humans.

The fact that in Eurasia and Africa the phenomenon had smaller dimensions suggests that where humans evolved their hunting skills, the animal species learned to fear them and to survive their hunting, while where they arrived suddenly the animals had no time to develop a suitable behavior and became extinct. This leads us to doubt the correctness of the often encountered statement that intelligence has little survival value for a species; its value seems to be even too large, in the sense that the intelligent species tends to destroy the other species and replace them, thus becoming another cause of mass extinction—at least until the time it understands the dangers implied by such behavior.

CONDITIONS NEEDED FOR THE DEVELOPMENT OF LIFE

In the previous section the development of life on our planet was briefly dealt with, from the first unicellular organisms to the Cambrian explosion, to the spread on dry land of a myriad vegetable and animal species, and to the eventual creation of a complex biosphere that covers the whole planet. It is natural to wonder whether what happened on Earth may be representative of a general trend toward life or is a very rare occurrence and our planet is instead an island of life in a sterile Universe.

The speed with which life began and the variety of environments that are able to sustain it may lead us to support the first hypothesis. Yet between the idea of a sterile universe and that of a universe teeming with life, an intermediate hypothesis that is gaining support is known as the *rare Earth hypothesis*. Its premise is that life is an almost immediate consequence of the formation of planets and that elementary life-forms are therefore extremely common in the Universe. They may be very different from those we know on Earth and might use the most disparate forms of energy. When we make contact with them, it might be difficult to realize that we are facing living beings. These forms of life would be extremely resilient to planetary catastrophes, to the point of surviving immediately after the formation of a planet, when the meteoric bombardment is still very strong and the structure of the planet not yet settled.

However, the time that was needed to develop more complex forms of life on Earth and, above all, to reach the stage of animal life, may suggest that animal (complex) life is extremely rare and that Earth might be the only planet, at least in this part of our galaxy, to host it. This hypothesis is strengthened by the consideration that animal life is much more delicate and more prone to be canceled by catastrophes of various kinds. And if higher life is so rare, intelligent life must be even rarer.

Whether or not this hypothesis is correct, it has the merit of separating the conditions needed for life in its more elementary forms from those necessary for the formation of complex life. We should not speak, for instance, of the habitable zone in a planetary system, but of two zones, a larger one habitable for bacteria and simple forms of life, and another one, perhaps extremely small, habitable for complex organisms and animal life.

The basic conditions for planets to be habitable come from astronomical considerations, and are determined by the planet's position in space. The position in the galaxy of the star around which the planet orbits must be considered first. A hypothesis has been formulated relating to the so-called galactic zone of intelligent life: it tends to restrict the possibility of life to limited areas of the galaxy.

The regions close to the nucleus are probably not habitable. The stars are much closer to each other than where our Solar System is located, and close encounters between stars must be more frequent. When this happens, the orbits of planets are perturbed and in the most extreme cases some planets may be expelled from the system or set on collision courses with a star. Even if these extreme occurrences are rare, the orbits may change in such a way as to produce changes of the surface and atmospheric conditions of the planets.

A high star density also increases the probability of stellar explosions close enough to cause serious damage to the biosphere of planets. Further, high-energy particles coming from the center of the galaxy may cause mass

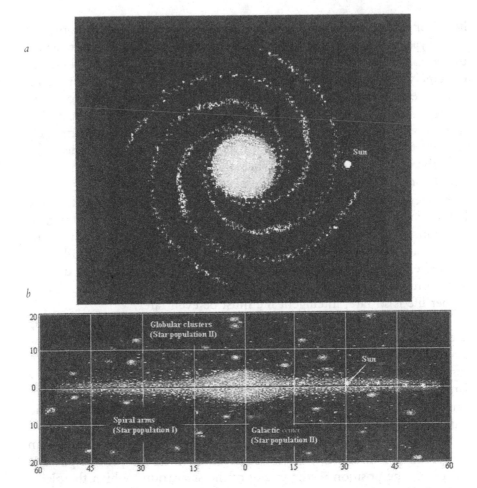

a

b

FIGURE 3.14 *The Milky Way; position of the Sun with respect to the spiral arms: (a) plan view; (b) cross-section. Scale in thousands of light-years (note that the scale in a direction perpendicular to the galactic plane is expanded).*

extinctions in nearby systems. A sketch of our galaxy, the Milky Way, is presented in Figure 3.14. The rather peripheral position occupied by the Sun is clear.

Not only the zones close to the nucleus but also some regions located in a more peripheral position can be dangerous to life. Our galaxy is a spiral galaxy, with arms in which the density of stars is higher than that in the zones between the arms. What has been said for the galactic center, may hold, albeit more weakly, for the arms of the spiral.

The problem here is that the galaxy does not rotate like a rigid body, and the speed of rotation of the stars is different from that of the form of the galaxy. This means that in the zones where the stars rotate faster than the arms or vice versa, the stars periodically get in and out of the various arms. The arms must not be thought of as material objects but simply as zones where stars are more dense. They are like pressure waves propagating in an organ pipe. The pressure waves move at a different speed from the molecules of air; therefore, every molecule moves continuously between zones of high and low pressure. Similarly, the stars move from zones of high density to zones of low density.

The speed of the stars and that of the arms vary with the distance from the nucleus, so that stars move faster than the arms near the center but more slowly than the arms in the outer parts of the galaxy. This means that there must be an intermediate zone, called the corotation zone, in which the two speeds are equal. The stars located in that zone and far from the arms are always in a good location for the development of life. Outside the corotation zone, stars periodically enter into the arms and, if it is true that life is not possible there, then life in such systems is periodically canceled. If this happens frequently enough, there would not be time enough for complex life to evolve.

The habitable galactic zone, therefore, would be limited to the parts of the corotation zone that are between the arms. This theory is still hypothetical, since it is not known whether life in the arms of the spiral is indeed impossible. Moreover, there are not enough data to calculate the exact position of the corotation zone and only rough estimates exist. Yet the Sun is located far from the arms and its distance from the galactic center coincides almost exactly with the estimated position of the center of the corotation zone.

The type of star around which the planet orbits would also seem to be important for the possibility of developing life. Stars that are too small, for instance, are also colder and the planet must then be close to the star in order to have a high enough temperature to host life. But if a planet is located very close to a star, it has a tendency to remain gravitationally locked, with a speed of rotation coincident with the speed of revolution, as is the case with the Moon, which always shows Earth the same side. But in this way one hemisphere of the planet will be extremely hot while the other will be very cold, and this makes it very difficult for life to start.

In the case of very large stars, the problems are not tied to the planet's distance from them, but to the fact that large stars burn their nuclear fuel at a fast pace and their life span is much shorter than that of our Sun. The life of a giant star is so short that it is doubtful that complex life can develop on its planets before they are charred by the star, which, like all big stars, ends its life as a nova. It has been calculated that a star must be larger than half the size of the Sun, but smaller than twice its size, to have planets that can host life.

Life on Earth is based on chemical reactions taking place in water solution. Reactions occurring in the solid phase are probably too slow, and those in the gaseous phase too violent to allow for the existence of forms of life not based on a solvent of some kind. But must this solvent be water, or may life be based on other liquids? Certainly water has some very particular properties, like having a smaller density in its solid phase (ice) than as a liquid. This is undoubtedly an extremely useful property; in fact, it is quite likely that without it life on Earth would not be possible, but is it absolutely essential? Assuming that life must be based on carbon, could life exist on planets colder than Earth that use ammonia or hydrocarbons like methane as a solvent?

Not knowing whether forms of life based on solvents other than water are possible, life is assumed to exist only on planets whose surface temperature allows water to be in the liquid state. The temperature must therefore be between 0°C and 100°C—assuming that the atmospheric pressure is approximately equal to that on the surface of Earth. If the pressure is high enough, water remains liquid up to higher temperatures, as in the case of deep submarine hot vents, while if the pressure is very low, ice sublimes without passing through the liquid phase.

In the past, the distance from the star was assumed to be the main feature that determined whether life could exist on a planet, to the point of defining a habitable zone for every system. Yet there is no agreement on the extension of such a habitable zone; in the case of the Solar System some think that only a narrow ring, including just the orbit of Earth, is habitable, while others assert that Venus and Mars are also included.

Moreover, the surface temperature of a planet depends not only on its distance from the star but also on the composition and the pressure of its atmosphere. An atmosphere made mainly of diatomic gases such as oxygen or nitrogen has a very weak greenhouse effect, allowing a large part of the heat reflected by the surface to be lost in space. Triatomic gases like carbon dioxide or water vapor or, even more so, complex molecules like ammonia, methane, and chlorofluorocarbons (CFCs, which, however, do not exist naturally in planetary atmospheres), cause a strong greenhouse effect, producing much higher surface temperatures. Earth, for instance, would probably be too cold were it not for carbon dioxide and water vapor: the average temperature of the Moon, which is at the same distance from the Sun as Earth, is about −18°C. Without a greenhouse effect, Earth (or at least large parts of its surface) would probably be unsuitable for life, like Mars. Further, the composition and density of the atmosphere depend on the composition of the surface and subsurface layers, in a complex dynamic equilibrium. Lastly, the ability of the surface to reflect thermal energy back into space has a large influence on its temperature.

An example of a planet too hot for life is Venus; true, it is closer to the Sun than Earth, but the enormous difference of temperature cannot be explained by this circumstance alone. In the past it was thought that the habitable zone of the Solar System was outside its orbit, but today we are no longer sure that such simple explanations are viable. The present environment on Venus seems the product of a runaway greenhouse effect; an increase in the temperature of the planet caused the evaporation of the seas, if they ever existed on Venus, and above all the release of carbon dioxide from the carbonates in the crust.

The increase of water vapor and carbon dioxide in the atmosphere caused a further increase in the greenhouse effect and therefore further heating. At the end the seas boiled away and all the carbon dioxide contained in the ground was liberated into the atmosphere, causing the dense and hot atmosphere we find today. Water vapor was decomposed by sunlight into oxygen and hydrogen, and hydrogen was lost into space. So it is not the case that Venus contains more carbon dioxide than Earth, only that it is all in the atmosphere instead of being in the ground and in the seas. Venus is really a red-hot hell, with temperatures of about 500°C and with an atmospheric pressure on the surface roughly 90 times that on the surface of Earth.

The opposite example among Earth's neighbors is Mars. Currently the atmosphere of the Red Planet is extremely thin: the pressure on the surface is about one hundredth of the atmospheric pressure on Earth and the temperature recorded by the two *Viking* probes ranges from −14°C to −120°C. The atmosphere of Mars, like that of Venus, is mainly made of carbon dioxide, but in this case the density is too low and the greenhouse effect is not strong enough to prevent the surface temperature from being very low. Liquid water is not stable on the surface of Mars; the South Pole cap is partly made of water ice (the greater part of it, as well as the cap on the North Pole, are made of carbon dioxide frost, or so-called dry ice), which evaporates directly when heated by the Sun.

Mars, however, has not always been like it is today. In the past—a remote past, perhaps more than three billion years ago—it had a thicker atmosphere, with a density perhaps twice that of Earth's atmosphere, and the surface was much warmer, to the point that evidence has been found of the presence of water flowing on its surface. Mars probably cooled down owing to a process exactly opposite to that heating Venus; the cooling of the surface fixed much of the atmospheric carbon dioxide into the ground in the form of carbonates, reducing the greenhouse effect and therefore causing further cooling. Mars probably does not contain less carbon dioxide than Venus today, but it is almost all in the ground instead of being in the atmosphere. Mars is considered too cold to host life, and therefore the habitable zone of the Solar System includes neither Venus nor Mars. However, the habitable zone might be much wider in the presence of a planet with a suitable atmosphere.

All stars, including the Sun, become warmer as they get older and thus the habitable zone moves outward with time. Perhaps Venus was once habitable and, if life actually develops quickly when it has the opportunity to do so, inhabited by very simple forms of life. Then the increase in temperature caused by the heating of the Sun may have started the runaway greenhouse effect, leading to the present situation.

In the past there were scientists who thought that Mars, in a similar way, would be habitable in the future, when the Sun will be warmer, but this is contradicted by the observation that Mars was probably habitable in the past. In this case it is possible that the cooling that started the process leading to the loss of the planet's atmosphere, with the subsequent decrease in greenhouse effect, was caused by a catastrophic event, like an asteroid strike. Though some scientists claim that this event occurred rather recently, at least in geologic terms, it is very likely that Mars cooled relatively soon after its formation, perhaps when the heat emitted by its radioactive elements decreased. Relatively warm conditions seem to be unstable on small planets, for they tend to lose their atmosphere, hence reducing the greenhouse effect.

As already said, the temperature of a planet depends mainly on the presence of greenhouse gases in its atmosphere, mainly carbon dioxide. To maintain a temperature suitable for life, the carbon dioxide cycle between the atmosphere and the crust of the planet is therefore essential. On Earth the mechanisms of plate tectonics seem to be fundamental for this goal: the crust of the planet is divided into zones, or "plates," floating on the external part of the underlying mantle and moving, slipping one close to another or one below another. Plate tectonics is responsible for the continental drift that modifies, extremely slowly but without rest, the form and size of the continental masses and the oceanic basins.

It was well known that plate tectonics had an influence on the evolution of life as the shape of the continental masses, their continuity—when there was just one continent, Pangaea (in the beginning of the Jurassic)—or their present fragmentation, contributed to create the environment that shaped evolution. If life began near underwater hot vents, the mid-ocean ridges where new crust is formed at the junctions between plates may have had an important role, though it is not quite sure that movements of the mantle started before the beginning of life. Rather, it now seems that plate tectonics started two or three billion years ago, and therefore could not have had a role in the beginning of life. Yet it is possibly linked to the generation of oxygen that allowed the development of animal life.

Today a much more important role has been attributed to plate tectonics: while causing new crust to be continuously created in the mid-ocean ridges, it also causes parts of the old oceanic crust to be carried again under the continental masses, where it re-melts to become magma. In this way part of the carbon dioxide fixed in the ground is once again liberated into the atmosphere to maintain the greenhouse effect. Another reason that could make plate tectonics necessary for life is linked to the magnetic field: a planet has an intense and long-lived magnetic field only if its nucleus remains in the melted state, and plate tectonics might be essential for this. In the absence of an intense magnetic field, life, or at least complex life, can be easily destroyed by radiation from space.

It would seem that a planet must display plate tectonics to maintain the conditions needed for life for a sufficiently long time to allow the development of its more complex forms. Neither Mars nor Venus, nor any other planet or satellite of the Solar System, presently has true plate tectonics, even if some traces suggest that Mars may have had such a feature in the past. Finally, a hypothesis was put forward that the body that is ultimately responsible for plate tectonics on Earth is its large satellite. As already mentioned, the circumstances of our satellite's formation are quite peculiar, so if this hypothesis should prove correct, planets with plate tectonics could be rare indeed, even outside the Solar System. And if plate tectonics is a necessary requisite for the development of advanced forms of life, complex life would be even rarer.

At the beginning of the 1970s James Lovelock and Lynn Margulis proposed the theory that life itself controls Earth's environment and keeps it suitable for life. This hypothesis was given the name of *Gaia*, the Earth goddess of the Greek–Latin pantheon. The hypothesis is based on the fact that many cycles regulating and keeping the temperature and composition of the atmosphere constant are based on the presence of living beings, as, for instance, at the simplest level, the oxygen–carbon dioxide cycle involving animals and plants. But the *Gaia* hypothesis goes well beyond this and claims that the cycle regulating the presence of carbon in the form of carbon dioxide in the atmosphere—and of carbonates in rocks and sea sediments—thus acting as a thermostat, is also governed by living beings.

Statements like these arouse much criticism, particularly when they are pushed close to (and even beyond) the limits of what resembles more a form of Earth-mysticism than a consistent scientific theory. It can be even obvious to state that the planet and its living beings evolved together and constitute a single integrated system, but it is difficult to see how living beings could undertake tasks on a planetary scale, like transforming the primitive atmosphere based on carbon dioxide into one based on oxygen, as if they had a precise project in their minds. Moreover, an oxygen-based atmosphere was harmful for the forms of life of that time and would only have become useful later, when new life-forms developed that were able to exploit the new oxidizing atmosphere. The stronger forms of the *Gaia* hypothesis see the planet as a single living unit, acting in such a way as to regulate its physical and chemical environment with the purpose of maintaining conditions suitable for its own life. Thus Earth's atmosphere appears to be not only an anomaly, when compared with the simple evolution of the planet, but like a tool created by the biosphere to fulfill its purposes. It must be concluded that the weaker forms of the *Gaia* hypothesis do not appear to add anything new to older theories, while the stronger forms, with their finalism, have little to do with science.

Up to now the habitable zones around the Sun (or any star) were defined according to the possibility that the surface temperature of a planet stabilizes enough to allow liquid water to exist. Yet a celestial body can be maintained at temperatures suitable for life by phenomena other than the heat from the star around which it orbits. Radioactive decay inside a planet may not be sufficient to maintain a planet's warmth for a long enough time to allow the development of complex life-forms, but tidal heating, such as occurs in the satellites of large planets like Jupiter or Saturn, can seemingly maintain a body at high temperatures. A typical example is Io, a satellite of Jupiter, rich in volcanic phenomena due to tidal effects. Another satellite of Jupiter, Europa, probably has a deep ocean of liquid water under its frozen surface and is a good candidate to host life. These satellites are well outside the Solar System's habitable zone (at least by its traditional definition), yet may have liquid water on their surface or below it.

Many arguments used to establish the possibility of life on a planet are based on the conditions on its surface, intending for "surface" the interface between the solid or liquid zone of the planet and the gas lying over it. But it has recently been realized that on Earth many living beings thrive in the depths of the ground and that life may even have started in such conditions. It is possible to think of the same thing happening on planets whose surface is not habitable because it is frozen, bombarded by cosmic radiations or exposed to the space vacuum, and therefore life can only exist in inner pockets where it can find liquid water and energy, even in the form of thermal gradients that can be exploited. Taking this line of reasoning further, even on planets whose surface is too hot, liquid water temperatures may be present at a certain height in the atmosphere, where microorganisms might exist, living in the droplets of water of the clouds.

In any case, all the considerations above are based on the assumption that the optimal conditions for life are those occurring on Earth and postulate that life must necessarily be of a terrestrial type. Water is certainly essential for life on Earth, and its properties as a solvent are, as already mentioned, precious for almost all the chemical reactions on which our biology is based. But if forms of life based on other solvents, such as methane, are conceivable, then should we not look for other habitable zones, based on the existence of, say, liquid methane?

For that matter, does the fact that radiation is harmful to living beings on Earth (and not for all in the same way, since some of them are particularly tolerant of radiation of every sort) allow us to rule out the existence of living beings thriving in the zones close to the center of the galaxy, in the spiral arms, or in other zones rich in radiation? For all we know, they may even use the energy of gamma radiation instead of sunlight!

Finally, two points made by proponents of the rare Earth hypothesis must be considered. The first concerns the Moon. It has already been said that the tides due to the combined action of the Sun and the Moon had an important role in the birth of life on Earth. But the Moon may also be important for another reason: the presence of such a large satellite has the effect of stabilizing the rotation of Earth around its axis. Without the Moon, Earth could have experienced large variations in the inclination of its rotational axis, that are extremely harmful to life. If the axis lay on the orbital plane (an inclination of 90 degrees, as in the case of Uranus) instead of being almost perpendicular to it (actually it is tilted at 23 degrees), no regularity in the day–night cycle would exist; months of illumination would alternate with months of darkness. The thermal excursions would be extreme and life on the planet would probably be impossible. Further, sudden variations in the inclination of the axis would cause cataclysms leading to the destruction of many ecosystems. A large satellite like the Moon guarantees the stability of the planet's conditions and the persistence of life.

The second consideration concerns the presence of Jupiter, that is, of a giant planet in a stable orbit external to the habitable zone. It has already been noted that a giant planet in a strongly elliptical orbit would probably render our system uninhabitable, as would a large planet on an orbit that decayed in time until it came close to the star. Both these occurrences have been observed in extrasolar planetary systems (in the latter case it has only been deduced that the planet came close to the star, since there is no theory explaining its formation there).

If a Jupiter in a "wrong" position is harmful, having a Jupiter in a "correct" position could be essential for the development of life, or at least of complex life. The orbital perturbations caused by Jupiter on the planetesimals orbiting in the zone immediately inside its orbit not only prevented the formation of a planet in the zone where the asteroids belt is found, but also subtracted material from Mars, which is much smaller than Venus and Earth, and, above all, it cleaned up, so to speak, the inner Solar System from potentially dangerous planetesimals.

Many of the planetesimals that would have fallen on Earth, even in recent times, were deflected to fall into the Sun, be expelled from the system, or deflected to collide with a planet in earlier times, reducing the probability of dangerous impacts. If the current probability of an asteroid of 10-km diameter (like the one that caused the K/T extinction) falling on Earth is one impact every 100 million years, G. Wetherill of the Carnegie Institute in Washington computed that without Jupiter in its place the frequency of impacts of that kind would be much higher—as high as one every 10,000 years.[14] The presence of Saturn, another giant planet, though smaller than Jupiter, in an external orbit has also been very beneficial.

These considerations tend to restrict the conditions for the habitability of a planet. Actually they seem to lead to the conclusion that only a planet orbiting around a star exactly like the Sun, in exactly the same position in a galaxy, exactly at the same distance from the star, with a satellite just like the Moon, in a system with a planet like Jupiter in that orbital space, and with plate tectonics as on Earth, may host life not limited to the simplest microorganisms. At this point a question comes naturally: doesn't this reasoning apply only to life of exactly the terrestrial type, thus demonstrating (the fairly obvious fact) that forms of life that developed on Earth could only have developed here, but saying nothing about other forms of life?

Actually, the basic problem of astrobiology is always the same: as someone ironically pointed out, it is the only science without an object to study. Or at least, since probably the object of astrobiology (extraterrestrial life) exists, it is the only science that cannot study its object and does not even know with certainty if it exists. As it waits for the possibility of studying extraterrestrial life, it assumes that it is similar to the terrestrial one, with the risk of discovering that it can exist only on Earth.

It is probably true that complex life is much more delicate than the

[14] P.D. Ward and D. Brownlee, *Rare Earth*, Springer, New York, 2000, p. 238.

bacteria and other life-forms that may develop quickly on many celestial bodies in conditions different from those on Earth, yet it is also likely that different forms of complex life could develop in different environments. Perhaps no terrestrial animal or plant could live on a planet close to one of the spiral arms of the galaxy, but on a planet that is subjected to strong radiation, life-forms could evolve that are more resistant to radiation, or can even use radiation from space as an energy source.

Science fiction accustomed us to the most disparate hypotheses on this subject, such as forms of complex life able to live in the extremely low temperatures of the planetoids of the Kuiper belt, in the atmosphere of the giant planets, in the intense gravitational fields of the neutron stars or in the vacuum of space. To live in these extreme conditions such forms of life would not only have to be different from terrestrial organisms, they must be based on a completely different biochemistry and, in certain cases, use matter in a different state from what we are accustomed to. But until we find evidence of their existence, we must consider them for what they actually are: fruits of our imagination and of our need not to be alone.

LIFE ON MARS

The only bioastronomic research ever carried out on another planet was performed by the *Viking* probes on Mars (Figure 3.15), and it gave inconclusive and essentially negative results. A map of the planet is shown in Figure 3.16.

The successive hopes, certainties and disappointments related to the existence of a Martian civilization, or at least of animals and plants and, as a last resort, of microorganisms on the Red Planet were described in some detail in Chapter 1. The lowest point of these hopes was perhaps reached in 1966, with the photos taken by the *Mariner 4* probe; the desolate lunar landscape really looked like a hopeless world. It was essentially a stroke of bad luck, as if a hypothetical extraterrestrial, having the possibility of obtaining just a pair of images of our planet, got two photos of the most arid zones of the Arabic desert, in dry season! But nobody knew that then, and even when the *Mariner 9* images arrived, the expectation of finding life did not greatly improve.

Nevertheless, not the scientists who prepared the *Viking* mission did their best and included four biological experiments in the payload. The first experiment was by a *gas chromatograph–mass spectrometer* (GCMS), able to search for organic molecules in specimens of soil. The instrument could

FIGURE 3.15 Winter frost in the landing place of the Viking 2 probe in the Utopia Planitia. The ice is probably mostly frozen carbon dioxide (dry ice). (NASA photo.)

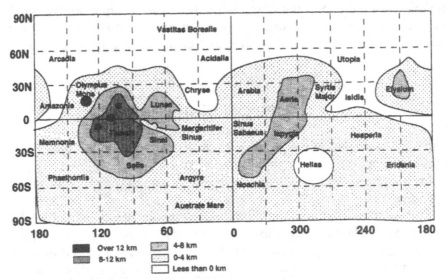

FIGURE 3.16 Schematic map of Mars (from R. Zubrin, The Case for Mars, Touchstone, New York, 1997).

distinguish between organic substances of biological and nonbiological origin.

A second experiment was aimed at identifying the products of photosynthesis reactions released by soil specimens (*pyrolitic release experiment*). Then there was an experiment in which specimens were bathed in a broth of nutrients favoring the proliferation of possible microorganisms (*gas-exchange experiment*). Finally, in the last experiment, the specimens were put in contact with organic nutrients marked with radioactive isotopes, in such a way that organisms coming in contact with them would develop gases containing the radioactive isotopes and therefore be easily identifiable (*labeled release experiment*).

When the first *Viking* lander finally settled on the surface of Mars, the impression that the planet was a desolate body like the Moon disappeared, leaving in its place the awareness that the probe was on a real planet, a real "place," as Carl Sagan said, noting that it was not much different from some places in Colorado or Arizona.

A little later the instruments began their work and the scientists announced that two of the experiments had given positive results. After dipping the specimens of soil in the nourishing solution, the gas-exchange experiment revealed the presence of oxygen and the labeled release experiment also gave positive results. Yet the scientific world reacted with skepticism to the announcement and wisely waited for the results of the other experiments.

When the results of the gas chromatograph–mass spectrometer arrived, they were a big disappointment for everyone: there were apparently no traces of organic substances of any sort on the surface of Mars. And the strange thing was that there weren't even those organic substances of non-biological origin that are so common on asteroids and comets (though in the 1970s this was not as certain as it is now) and that are present even on the Moon. Sagan himself was surprised: "If there is life on Mars, where are the dead bodies? . . . Martian soil has less organic matter than the surface of the Moon."

Some weeks later, when the second *Viking* lander arrived on Mars the experiments were repeated, yielding identical results. There could only be one explanation: the soil does not contain organic substances and there is no form of life. The ultraviolet and ionizing radiations, not filtered by a layer of ozone and a magnetic field as on Earth, sterilize the surface. The reactions that had, at first, suggested the release of gas from biological specimens must be ascribed to reactions between inorganic matter, mainly the strongly oxidizing substances present on the ground. Actually, some of the scientists involved, and particularly Gil Levin, responsible for the "labeled release experiment," claimed that the positive results were evidence of the presence of life, but most participants in the experiments drew negative conclusions, although admitting that no conclusive result could be drawn from the experiments. The absence of evidence is not equivalent to evidence of absence.

The two *Viking* astrobiological experiments were recently reinterpreted in a somewhat less negative light, and it is therefore impossible to rule out absolutely that some primitive form of life exists on Mars. Moreover, the fact that on Earth life occupied all available space and created an actual biosphere extending on the surface and under the ground for a depth of thousands of meters does not mean that on a planet where life constitutes more of an exception than the rule, it might not be limited to some small, well-defined zones. The landing spots of the *Vikings*, the *Pathfinder/Sojourner*, and the more recent probes were chosen to allow easy landings and are certainly not the most suitable to find forms of life. From this viewpoint, deep canyons like the Vallis Marineris, where the scarce atmospheric moisture could gather, would have been much better, but performing an automatic landing there is nearly impossible. The most suitable zones for life can be reached only by vehicles, piloted or automatic, moving on the ground and able to manage the steep slopes of the mountains.

The discovery of extremophile bacteria living deep underground occurred after the *Viking* experiments and opened up new possibilities. Even if the surface is sterilized by radiation, life might survive below it, perhaps in the permafrost layers that seem to extend from a depth of a few meters to tens of meters. The announcement, in August 1996, of the possible presence of fossils of very primitive life-forms in a meteorite, originating from Mars and found in Antarctica (Figure 3.17), made a big sensation and stirred arguments. The meteorite is of the SNC type, a well-known type to which the Shergotty meteorite found in India in 1865 belongs (they are also called Shergottites), as do meteorites found in Nakhla (Egypt) and Chassigny (France). The designation SNC derives from the initials of the three localities.

FIGURE 3.17 ALH84001 meteorite found in 1984 at Allan Hills in Antarctica. (NASA image.)

There is little doubt of the Martian origin of the SNC meteorites; if nothing else, the isotopic composition of the elements from which they are made constitutes very convincing evidence. There is also no doubt that fragments from the surface of a rather small planet like Mars are in fact projected into space by the impact of a large meteorite. The piece of stone found in Antarctica is therefore surely from Mars, like the other of the same type.

The analyses of ALH84001 showed that that rock is very ancient, having crystallized about 4.5 billion years ago, and that it suffered a strong impact around 500 million years ago. Water flowing in the rock left carbonate deposits between 3.6 and 1.8 billion years ago and the formations that someone today interprets as microfossils were formed roughly at the same time. The impact that sent the stone into space occurred around 16 million years ago and from that moment, for millions of years, the meteorite wandered in the Solar System. Finally, 13,000 years ago, it entered Earth's atmosphere and remained on the Antarctic ice until Roberta Score picked it up. Initially it was classified as a fragment of an asteroid, a diogenite, and was stored, together with other meteorites, in a deposit at the Johnson Space Center in Houston. It was only in 1993 that it was again studied and reclassified as an SNC meteorite.

Only about 15 meteorites are known to have come from Mars; as a consequence ALH84001 became a rare piece, worthy of detailed study. Moreover, David Mittlefehldt saw a reddish zone that suggested inclusions of carbonates, and nobody had ever found carbonates in Martian meteorites. The object became more and more interesting and a group of scientists, led by David McKay, obtained various specimens of the meteorite.

McKay, using a scanning electron microscope with a magnification of 30,000, saw strange formations that looked like microscopic worms (Figure 3.18). Undoubtedly, those formations looked like microfossils even if, as was immediately said, they were too small, smaller than similar terrestrial formations. In the following years, however, microfossils of similar size were found on Earth, too.

Other researchers found polycyclic aromatic hydrocarbons (PAHs) in the same meteorite, which are often associated with life but are also known to be produced by nonbiological reactions, and traces of an iron sulfide, which often has a biological origin. Yet even if contamination by biological material from the Earth is very unlikely, it is still uncertain whether evidence that life existed on Mars in the past has been obtained. The microscopic structures are very similar to microfossils, but none of the clues gives the certainty of a biological origin.

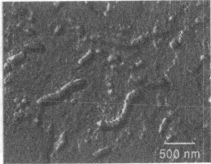

FIGURE 3.18 *Two images, at different magnification, of formations similar to microfossils in micro photographs of the ALH84001 meteorite. Their length is just 0.2 thousandth of a millimeter. (NASA photo.)*

Those who defend this thesis stress the fact that each clue is not proof in itself, but if they are taken all together they give a consistent picture.[15] On the other hand, it is stressed that putting many clues together does not constitute certain evidence.

Work on the Martian meteorite goes on and its scope has been widened to include other SNC meteorites, particularly those from Nakhla and Shergotty, in which formations of the same type were found. In this case what is more amazing is the age of the meteorites: the first is about 1.3–0.7 billion years old, while the second is far more recent, just 165 million years old. If these formations are really microfossils, Mars should not only have hosted living beings in a distant past, but even in relatively recent times, when dinosaurs were living on Earth.

If the discoveries on the SNC meteorites are indeed evidence of life, it would have started at the same time on Mars and on Earth—a period in which the Red Planet was rich in water and had an environment much different from the present one. The changes that occurred on the planet would seem to have aborted it in its initial stage, before it developed and occupied the entire planet. If, however, the 1999 discoveries are also confirmed, then we must conclude that it still existed in much more recent times, and may still exist even today.

Some scientists think that if life really started on that planet, it could have survived in some particularly suitable places, such as the permafrost that is likely to be present in the subsoil of much of the planet. If life still exists on Mars, however, it must be a very marginal component of the

---

[15] For a detailed discussion, see, for instance, Bruce Jakosky, *The Search for Life on Other Planets*, Cambridge University Press, Cambridge, 1998; and Laurence Bergreen, *The Quest for Mars*, HarperCollins, London, 2000.

planet, or at least of its surface, with large zones in which no trace of it can be found. The possibility of finding life on the Red Planet cannot therefore be completely ruled out and constitutes an incentive to continue exploration. The whole matter of the Martian meteorites, both in the case that they contain the first forms of alien life ever discovered, and in the more likely case that it is just a false alarm, must teach us two things.

The first is that the possibility of an exchange of matter between planets in the same system, but perhaps also in different systems, although unlikely, must not be discarded, and this would make the propagation of life from one planet to another a possible scenario. If there is (or there has been) life on Mars, it is not impossible that life on two planets so close to each other has a common origin. The passage of material from Mars to Earth is easier than the other way around, owing to the weaker gravitational field of Mars compared with Earth. In the case of a common origin, it is very likely that life started on Mars and later migrated to Earth.

The second consideration is that the identification of extraterrestrial life-forms may be very difficult: almost 10 years have passed since the first announcement of the Martian meteorite, and although studies of all types have been performed in the best laboratories, there is not, as yet, a definitive answer. How could an automatic probe, able to perform only a few experiments and planned years in advance, supply certain evidence? If clear evidence of the existence of such life-forms should be obtained, the problem will be solved, but in these conditions a negative result cannot give any certainty.

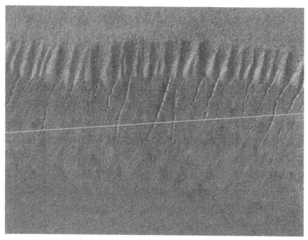

FIGURE 3.19 *Formations produced by water on a slope in the south pole area of Mars. (NASA photo taken by the Mars Orbiting Camera of the Mars Global Surveyor.)*

A mission in which soil specimens are carried back to Earth may give better results, but many of the problems in the study of ALH84001 will be encountered again; after all, it does not make a large difference if the specimen arrived here owing to natural causes or on board a space vehicle. The only advantages will be that the choice of specimen will be deliberate (but certainly not made in optimal conditions), instead of being left to chance, and that the trip will cause less damage to the specimen (launch, reentry on Earth, and exposure for a much shorter time to the space environment).

Before a negative answer can be given, a detailed study of different places on the planet and, above all, of the subsoil—where life could have formed, and could even still survive in the permafrost or in pockets in the rocks—must be performed. A program of this kind can keep the biological section of a scientific base on Mars busy for years, and perhaps even then the absolute certainty that Mars is a completely sterile planet will never be reached.

Some photos taken by the Mars Orbiting Camera (MOC) of the *Mars Global Surveyor* (MGS) have shown slopes carved by water streams (Figure 3.19). These formations are found in many places, but above all at high latitude, up to 70 degrees, not far from the south pole of the planet. These formations look geologically recent, and raise the possibility that liquid water existed on the planet in more recent times than normally thought. A few years ago it was suggested that temporary water courses can be formed on Mars when a large meteorite strikes the surface, melting large masses of permafrost. Hot water can actually flow for some time on the surface before evaporating and, at least in part, freezing. But it does not seem that in recent photos like that in Figure 3.19 anything of the kind happened. The hypothesis has been formulated that water very rich in salts may remain liquid on the surface of Mars for fairly long periods, at least in the warmest part of the day.

An even more interesting hypothesis is that Mars undergoes quick and radical climate changes, alternating cold, dry periods like the present one to warmer and moderately wet periods within a span of a few centuries. During the warm periods the evaporation of water in the polar ice caps could raise atmospheric pressure from the present 0.006 of Earth's atmospheric pressure to 0.03 or 0.04, enough to allow for liquid water on the surface. Some observations seem to suggest that Mars is presently warming up, like Earth. If this were true, the cause could be an increase the Sun's activity and could signal the start of a wet period on Mars.

The instruments of the *Mars Odyssey* probe, designed to detect water under the surface, detected, at least in the southern hemisphere, the possible existence of large permafrost layers just 1 meter under the surface. In these conditions, the possibility that life existed on Mars in the past, or may still exist, is strengthened. Some other clues have been found. A number of images taken by the MOC of the MGS show a large number of spots appearing on the dark dune fields in the southern polar region at the end of winter. These spots show a strongly seasonal behavior, both in their color and in their shape. The most accredited explanation is based on physical processes linked with the frosting and defrosting of the sand dunes, but alternative solutions based on photoautotrophic surface microorganisms have also been suggested.

The MARSIS (Mars Advanced Radar for Subsurface and Ionosphere Sounding) of the *Mars Express* orbiter started to look for underground water in July 2005, finding what looks like a huge (250 km diameter) impact crater, buried 2000 m below the surface in what appears to be material very rich in ice. This is surprising, since the place in which the buried crater lies—Chryse Planitia, the landing zone of one of the *Viking* probes—is a temperate, mid-latitude, location where no ice is found on the surface. It may mean that permafrost is present at warm latitudes, too.

The *Mars Express* found that Mars's atmosphere is richer in methane than expected, particularly in some areas such as over Arabia Terra, Elysium Planum, and Arcadia-Memnonia. This finding, quantitatively estimated at 10 parts per billion, was confirmed by observations with Earth-based telescopes. To maintain such a concentration of methane, this gas must be continuously produced on the surface or, better, in the subsurface layers, of the planet. Some researchers suggest that there are methanogen bacteria at work in these areas. This, however, did not convince the whole scientific community, since there are geological processes that can produce the same quantities of methane without the need for living organisms. It is a matter warranting further study, and one suggestion is to look for other hydrocarbons, like ethane: if the methane has a biological origin they should be absent, while their presence would support a geological origin. These clues of the possible presence of life make the contamination problem more delicate, both of the Martian environment by the probes and of Earth by specimens returning to our planet. The problem of contamination must of course be solved before astronauts can be sent to the Red Planet.

The probes are carefully sterilized and travel for months in a space environment. We know that this may not be enough and it is necessary to adopt very accurate sterilization procedures. Recently, a lively debate started on the possible contamination of Earth, particularly when it was decided to perform a Mars sample return mission in 1994. One of the two extreme positions is that of Robert Zubrin, clearly detailed in his various books.[16] Zubrin holds that there is absolutely no danger, because any parasitic organism evolved to infect a well-defined type of life (to use his words, "humans do not catch Dutch elm disease and trees do not catch colds"). Moreover, a Martian living being, provided such a thing exists, could not compete on Earth with beings that spent million of years adapting to the terrestrial environment. Besides, Zubrin says, Martian organisms have been transported to Earth by meteorites many times in the past without causing problems. He concludes that those who are afraid of Martian contamination had better leave Earth immediately!

The opposing view is that of the ICAMSR (International Committee Against Mars Sample Return), a body that, although not opposing sample return missions in principle, requests that the samples not be brought directly to Earth, but instead be left for a long quarantine on the *International Space Station*, where they can be studied and perhaps stored forever. The committee holds that Mars is almost certainly inhabited by microorganisms that, like all bacteria, are potentially dangerous for any form of life. The exchanges of biological material, even if they happen naturally through meteorites, are extremely dangerous. They hold that many epidemic diseases are potentially caused by meteorites, comets, and asteroids, in a position quite similar to Hoyle's version of panspermia.

Between these two positions, the majority of scientists and the NASA administration think that reasonable quarantine measures for all specimens originating from Mars must be taken. Zubrin's arguments seem to be sound, but, on the other hand, science cannot at present absolutely rule out the possible dangerousness of Martian bacteria, provided that they exist, and it is at any rate a wise policy to take precautionary measures even if the risk is very low. The outside of the reentry capsule with the specimens reaches very high temperatures, which guarantee its sterilization, and it is not difficult to design it in such a way that it arrives intact on Earth. There are many biological laboratories capable of opening the capsule and studying possible microorganisms in the required safety conditions, since there is long experience of dealing with pathogenic agents of every kind and virulence, including those artificially created for bacteriological warfare. Instead, it would be much more dangerous to study the specimens in a space station, owing to the difficulty of creating a sterile laboratory there. If there is no danger of contamination, the whole matter makes no sense; if something may be contaminated, the worst

---

[16] R. Zubrin and R. Wagner, *The Case for Mars*, Touchstone, New York, 1997; and R. Zubrin, *Entering Space*, Tarcher/Putnam, New York, 2000.

thing would be to have to deal with a contaminated space station and astronauts nobody knows how to bring back safely.

LIFE IN THE SOLAR SYSTEM

Mars is not the only candidate in the Solar System to host forms of life. Starting from Earth, the nearest body is the Moon. It is almost a commonplace that the Moon is sterile; all the analyses performed on specimens from our satellite confirmed this circumstance. Yet, at the time of the *Apollo* program, rigid quarantine procedures were still designed and applied to avoid contamination and will probably continue to be used if and when humans return from the Moon. The only further study that might be worthwhile, now that we know of the existence of extremophile bacteria living underground, is perhaps taking samples at various depths under the Moon's surface, but it is probably an excessive precaution.

The inner planets (Mercury and Venus) can hardly be considered in the search for living beings, at least as far as life similar to terrestrial forms is concerned. Mercury is similar to the Moon, both in the composition of its surface and in the absence of an atmosphere, but the proximity to the Sun causes much higher temperatures and radiation levels. Even if it is not gravitationally locked and does not therefore always expose the same side to the Sun, it does rotate so slowly that the day–night cycle is very slow. This could be even worse than being gravitationally locked, since if a side of Mercury were always exposed to the Sun, the other side could be rich in ice and an intermediate temperate zone would exist where life could develop. However, it is well known that water ice is present in Mercury's polar regions; it was this discovery that suggested that ice might exist in the polar regions of the Moon, too, a possibility then confirmed by robotic probes.

As already stated, the only possibility for forms of life based on water to exist on Venus is that they developed in the layers of the high atmosphere (around 50 km of altitude) where the temperature is not prohibitive. However, there is no clue that layers rich enough in water exist at all in the atmosphere of Venus. If Venus once had a lower temperature and atmospheric pressure, it is not impossible that life began there, too. In this case it likely ended when conditions started deteriorating toward its present state. However, we cannot rule out that some life-form could adapt to the conditions existing in the high atmosphere.

Mars has already been dealt with in detail. The outer planets, except Pluto, are all giant planets and although various ideas exist of what living beings fit for such environments might be like, they are still just hypotheses. Jupiter has an atmosphere made mainly of hydrogen and helium, with clouds of ammonia and some water, in the form of ice crystals or drops of liquid. The giant planets probably have a solid core, perhaps of hydrogen, around which a layer of liquid hydrogen exists. If the outer layers of their atmospheres are very cold, with increasingly low temperatures as the distance from the Sun grows, temperatures probably increase inside the planets. The possibility that a solid core exists depends on the temperature, pressure, and chemical composition of the various layers. Temperature and pressure, however, are such that the behavior of matter is rather different from what it is possible to study in laboratories on Earth.

There is surely a part of the atmosphere where the temperature is such that liquid water can exist and traces of water were found during the spectroscopic studies performed when the Shoemaker-Levy comet collided with Jupiter. It has often been suggested that in those layers of the atmosphere of the giant planets some living beings might have evolved, but again these are just hypotheses. No doubt, the substances used in experiments to replicate the formation of life on Earth—hydrogen, methane, ammonia, and water—are present and there is also plenty of usable energy. But any being that evolved in the atmosphere of Jupiter or Saturn, for instance, would have to avoid falling into the depths of the planet, where the pressure and temperature would destroy any organic substance. Living beings capable of floating, using balloons full of gas, or to fly aerodynamically, have been described, but it is difficult to imagine how they could evolve.

If living beings exist in the outer Solar System, it is more likely that they live on some of the satellites of the giant planets. Europa, a satellite of Jupiter and slightly smaller than the Moon, with a radius of 1565 km, is a possible candidate. Europa is covered by a layer of water ice that reminds us of our polar ice-pack and contains large quantities of water, probably 15 percent or more of the total mass of the satellite. The high-resolution image in Figure 3.20 shows a zone of 34×42 km of its surface: the ice is clearly fractured in a way suggesting that some plates, up to 13 km wide, were moving, floating on an underlying layer of water or more plastic ice, as happens with glaciers on Earth. Studies performed using radar altimeters and measuring the magnetic and gravitational fields also suggest that there is an ocean of water under the layer of ice, although neither the thickness of the ice nor the depth of the ocean is known.

A schematic cross-section of Europa is shown in Figure 3.21; the inner

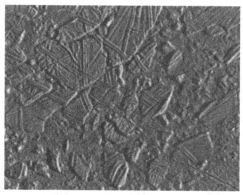

FIGURE 3.20 High-resolution image of a zone of 34 × 42 km of the frozen crust of Europa, a satellite of Jupiter, taken by the Galileo probe on February 20, 1997, from a distance of 5340 km. (NASA photo.)

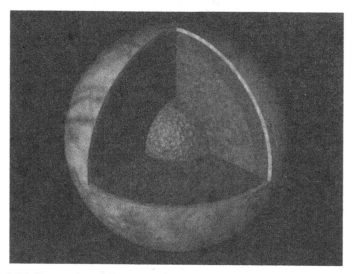

FIGURE 3.21 Cross-section of Europa, with a sketch of its possible internal structure. For the drawing of the surface, the images taken in 1979 by the Voyager probe have been used. (NASA image.)

details have been drawn using the gravitational and magnetic field measurements obtained by the *Galileo* probe. The core (not sectioned in the figure) is surrounded by a shell of rock (sectioned), which is in turn surrounded by a layer (darker in the figure) of water in liquid form or ice. The external layer, drawn in white, is surely ice. The ocean of liquid water, perhaps hundreds of kilometers deep, that may be present is then surrounded by a layer of ice whose thickness may be some kilometers, even 10 or more. In spite of its distance from the Sun, Europa is heated by tidal effects caused by the proximity of Jupiter. Certainly if the liquid ocean exists, it is possible that forms of life developed in it. Energy could be supplied by the thermal gradients in the water or by warm springs on the ocean floor and the surface layers of ice and water could protect living organisms from cosmic radiation and solar wind.

Europa is not the only satellite of the giant planets to be rich in water. Another Galilean satellite of Jupiter is Io: closer to Jupiter than Europa, it is strongly heated by tidal effects, to the point that it is the celestial body with the strongest active volcanism in the entire Solar System. Its density is very similar to that of the Moon, and it is probably mainly constituted of silicates. Apparently about 15 percent of Io is made of water ice, and the presence of hot parts suggests that zones where liquid water can form may exist on the surface or underground. Owing to the absence of an atmosphere, water should evaporate quickly on the surface, but underground lakes, where life might thrive, cannot be excluded. Protection from the strong radiations due to Jupiter is needed: Io's orbit is actually in the magnetosphere of the planet.

Titan, a satellite of Saturn, is more interesting from an astrobiological point of view. It is large, with a diameter of 5150 km, and one of the few minor bodies of the Solar System to possess a rather dense atmosphere, twice as dense as that of Earth, which hides the surface from direct observation. The atmosphere of Titan was studied in 1982 by the *Voyager* probe; then in 2005 the landing module *Huygens* of the *Cassini* probe reached the surface, taking pictures during the descent phase and after landing (Figure 3.22). The low density of Titan suggests that large quantities of water are present but it is likely to be in the form of ice, owing to the low temperature. At Saturn's distance from the Sun the surface temperature should be lower than 100 K (−173°C); the strong greenhouse effect due to the atmosphere raises the temperature somewhat, but according to measurements taken by *Voyager* it does not exceed −100°C, even at high altitude.

The atmosphere is mainly made of nitrogen, methane, and traces of argon; it is interesting mainly for the fact that it is very similar to the atmosphere that is thought to have existed on Earth early in its life and probably gave way to many experiments in organic chemistry. If this were true, the study of Titan could be very interesting for astrobiology, as it could allow us to observe the processes that started life on Earth.

Before the *Huygens* pictures were taken, lakes of liquid methane, or perhaps of methane and ethane, had already been supposed to exist on the surface, with the evaporation of methane, its condensation in clouds in the atmosphere and return to the ground as rain—an equivalent of the cycle

FIGURE 3.22 (Above) Composite image of Titan taken by Huygens during descent. (Right) The first image of Titan taken after landing on February 14, 2005. (NASA image.)

that water has on Earth. These ideas where confirmed by the images, in which a coastline seems to be clearly visible, together with clouds. If forms of life based on methane instead of water exist, Titan could be a good candidate to host them.

Beyond the giant planets and their satellites there is Pluto, with its satellite Charon, and the planetoids of the Kuiper belt. They are frozen worlds, at very low temperatures (a few degrees above absolute zero), rich in ice, some of it perhaps of water but above all of other substances that, in the rest of the Solar System, are found mostly in gaseous form (methane, ammonia, light hydrocarbons, etc.). It is possible to imagine life-forms suited to environments of this kind,[17] but they would be rather different from anything familiar to us.

Traditional astronomy, either optical or radioastronomy, can do little more than what has been done up to now to search for life in the Solar System. Even with extremely powerful instruments, placed in space like the *Hubble Space Telescope*, or on Earth, like the *Keck Telescope* with its 10-m-diameter mirror, one can discover little more than what we already know on the planets, their satellites, and the asteroids, particularly as far as the possibilities of the existence of life are concerned.

[17] R. Forward, "Alien Life Between Here and the Stars," in S. Schmidt and R. Zubrin, *Islands in the Sky*, Wiley, New York, 1996.

Interesting information can be obtained from the study of the fragments of celestial bodies that come to us: the meteorites. If indeed the biological origin of the organic substances in these space rocks can be confirmed or, better still, evidence can be obtained that the structures similar to microfossils are actually microfossils, a fundamental step will have been taken. Yet the difficulties are manifold. First, unless forms of life are found so clearly alien as to dissipate any doubt, it will be difficult to reach the absolute certainty that it is not contamination from terrestrial life. The experience of the last years has taught us that microscopic structures that can be interpreted as traces of life often allow different interpretations—it can be extremely difficult to reach definitive conclusions.

Only studies performed in place by automatic probes or by actual scientists, or performed in terrestrial laboratories on specimens collected by space vehicles, can give unequivocal results. Some missions are currently in progress: the *Stardust* probe, for instance, approached the comet P/Wild 2 on January 2, 2004, as close as 236 km, collecting thousands of particles of cometary and interstellar dust that have been brought back to Earth, and taking very interesting images of the comet. In particular they show that its surface is solid and not a rubble pile as was expected. The probe survived the close encounter and landed in Salt Lake Desert in Utah on January 15, 2006. The particles' size is between 0.1 and 0.01 mm.

A similar mission is the Japanese *Hayabusa* (formerly designated as Muses-C), a solar-electric spacecraft, launched on May 9, 2003. On November 26, 2005, the second attempt at landing on asteroid Itokawa succeeded, and on November 28 it was confirmed that the probe had collected about 3 grams of dust and bits of rock. The plan was to bring the samples back to our planet, but it now seems that this last part of the mission will fail.

Another mission that was initially aimed at bringing cometary samples to Earth was the European *Rosetta*. After being downsized several times, it was launched on March 2, 2004, with the aim of reaching the comet 67P Churyanov–Gerasimenko after a trip lasting eight years, during which it will pass close to two main belt asteroids, Steins and Lutetia. The landing of the *Philae* module and two years of studies of the comet will follow, and then the probe will approach the nucleus at 1 km. On June 14, 2005, *Rosetta* was 46.5 million km from the Earth.

The *Cassini–Huygens* NASA/European Space Agency (ESA) probe reached the Saturn system after having performed the flyby of Venus, Earth, and Jupiter The main vehicle, the *Cassini* probe, will remain in the Saturn system, flying close to the many satellites and sending a great deal of data and photos to Earth. The lander *Huygens*, with the help of a parachute, entered the atmosphere of Titan and, as mentioned earlier, sent back pictures of the surface of Saturn's principal satellite for the first time. In 2005 other probes were still active or completing their missions, like the *Mars Global Observer*, *Mars Odyssey*, *Deep Space 1*, and the rovers *Spirit* and *Opportunity*, while others were in the preparation stage. Their aim is the exploration of Mars, Venus, Mercury, and Pluto. None of them, however, carries experiments specifically designed for astrobiological work.

One doubt, however, remains: will progress in the field of artificial intelligence and robotics allow us to build probes able to perform automatically the delicate studies needed to recognize life-forms, possibly very different from those to which we are accustomed, in a near future? If, as it is fair to suspect, it should turn out that this is not feasible, it will be necessary to consider the human exploration not only of Mars but of other celestial bodies. Robots will certainly play an important role, particularly in hostile environments like those of the giant planets' satellites, but only the presence of humans can make the difference in bioastronomic studies—assuming that human scientists will be able to recognize forms of really alien life.

THE SEARCH FOR LIFE OUTSIDE THE SOLAR SYSTEM

As we saw earlier, after the first discovery of an extrasolar planet in 1996, other discoveries occurred at a rapid pace and a large number of extrasolar planets has now been discovered. However, the techniques currently employed only permit us to discover giant planets orbiting close to their star, and above all do not allow us to know anything about their physical and chemical characteristics. From an astrobiological point of view, it is simply the confirmation of a possibility, and nothing more. Seeing extrasolar planets is extremely difficult, not only because very powerful telescopes are required owing to the enormous distance, but above all because planets are faint objects, completely surrounded by the light of their star. One must remember that even the most powerful telescopes are not able of see details of the stars, which appear like bright spots and not as disks.

Interferometric techniques make it possible to combine the light captured by two or more telescopes as if they were parts of a single, large instrument. In this way it will be possible in the future to obtain images of extrasolar planets and, through spectroscopic studies, to investigate their composition. High-resolution interferometers must be located outside Earth's atmosphere even if they operate in the field of visible light. The study of extrasolar planets will mostly be performed using infrared light, since at those wavelengths the light of the star is less predominant over that from its planets. Thus operating outside the atmosphere becomes a necessity, since infrared light cannot reach the surface of Earth.

The Moon is an ideal place to locate a large interferometer, but before the colonization of the Moon is undertaken, it will be possible to build an orbiting interferometer, constituted by a number of vehicles, each one carrying a telescope (Figure 3.23). In order for the instrument to work properly, the distance between the telescopes must be controlled with extreme precision, with errors smaller than the wavelengths of light (a

FIGURE 3.23 *The Darwin mission proposed by ESA: a number of telescopes constitutes a large interferometer set in space not far from Earth. The four telescopes and the central vehicle with the optics for the reconstruction of the image are represented. (Courtesy of ESA.)*

fraction of a thousandth of a millimeter), something at the limits of present technologies.

ESA selected the mission *IRSI* (InfraRed Space Interferometer) *Darwin* as a part of the *Horizons 2000* program. It is based on an interferometer made of four telescopes, each with a mirror of 1.5-m diameter, located at a distance of about 1.5 million km from Earth. In the version sketched in Figure 3.23, the four telescopes are carried by four different vehicles flying at a distance of 500 to 1000 m from each other, with a central vehicle containing the optics that combine the light coming from the various telescopes. This allows the light of the star to be canceled through destructive interference, so that the weak light of the planets can be seen and analyzed.

The positioning of the vehicles must be very accurate, with a precision of a millionth of a millimeter, which is achieved with small rockets. To obtain detailed images, different exposures performed with the mirrors in different relative positions are added, which requires an even more accurate positioning. The interferometer can also be used to observe the spectra of light coming from the planets, thus allowing us to study the composition of their atmosphere. The launch of the instrument is scheduled for 2015.

The black and white conversion of the simulated false-color image obtainable aiming the spectroscope toward the Solar System from a distance of 10 ps (33 light-years) and performing a 10-hour exposure is shown in Figure 3.24. The Sun is not visible (its image is canceled to allow the planets to be seen) and the three bright objects are Venus, Earth, and Mars. The position of the planets can be measured so that it is possible to follow them during their orbit, but no detail can be seen on their surface.

The spectrum of the dot corresponding to Earth in Figure 3.24 is presented in Figure 3.25. In this case a longer exposure, 40 days, was simulated. The quantity of light received is given in photons per pixel of image as a function of the wavelength. For a body at the temperature of Earth but without a similar atmosphere, the spectrum shown by a dashed line would be obtained. Since the various gases in the atmosphere absorb light at well-determined wavelengths, the spectrum related to a planet with an atmosphere like that of Earth (full line) shows a decrease of the light intensity corresponding to the absorbed wavelengths. From the figure, the presence of molecules of water (H_2O), ozone (O_3), and carbon dioxide (CO_2) can be deduced. The presence of water suggests that the planet could host life, while that of ozone is even more explicit: large quantities of ozone may be present in the atmosphere only if it

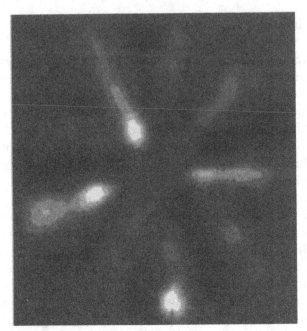

FIGURE 3.24 *Black and white conversion of the simulated image obtainable aiming the spectroscope toward the Solar System from a distance of 33 light-years and performing a 10-hour exposure. The three bright objects are Venus, Earth, and Mars.*

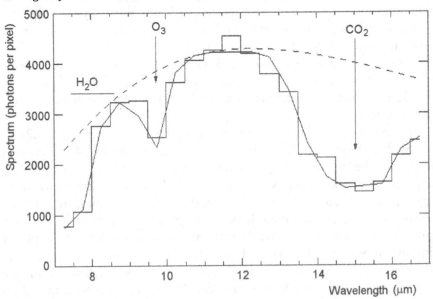

FIGURE 3.25 *Spectrum of the light of the point corresponding to the Earth in Figure 3.24. Dashed line: spectrum of a planet without atmosphere. Full line: spectrum of a planet with a terrestrial atmosphere. The absorption due to water vapor (H_2O), ozone (O_3), and carbon dioxide (CO_2) is visible.*

contains a high percentage of oxygen, a circumstance that indicates the probable presence of life in activity.

A NASA mission similar to *Darwin* is the *Terrestrial Planet Finder* (TPF): four telescopes with a diameter of 3.5 m, on four separate vehicles, plus a fifth platform containing the optics to reconstruct the image. The system should be positioned in the Lagrange point L2 of the Earth–Sun system. In this case the purpose is also to image extrasolar planets of terrestrial size up to distances of 50 light-years and to perform spectroscopic studies. The study of protoplanetary disks and of the formation of planetary systems is also planned. The TPF mission belongs to a wider program called *Origin*. Its purpose is to produce a catalogue of the terrestrial planets in the zone of the galaxy closest to us and analyze their atmospheres with the aim of identifying the habitable and the possibly inhabited ones. An initial schedule is to launch TPF in 2011.

A completely different approach to the search for terrestrial planets is to monitor the luminosity of the stars, in order to detect any decrease when a planet passes in front of the star (partial occultation). To observe the presence of terrestrial planets in this way, the luminosity must be measured in an extremely precise manner (some parts in 100,000). The measurements, therefore, must be performed in space, away from all possible disturbances. To be sure of the discovery of a planet, the occultation must be identified in a manner that can be repeated. The size of the planet can be assessed from the diminution of light intensity, while the period between two successive occultations allows us to obtain the data of the orbit and therefore the distance of the planet from the star.

An advantage of the method based on partial occultations of the star by the planet is that it can be applied even to very distant planetary systems. If the system, however, is sufficiently close to us it is possible, by studying the differential absorption of the star's light during the occultation, to get information on the composition of the planet's atmosphere.

The probability that a planet is in an orbit that leads it to pass exactly in front of the star, lined up in our direction, is very low, and therefore many thousands of stars must be continuously monitored to have a good chance of observing a certain number of events.

A mission of this type is *Kepler* by NASA, based on a custom-designed space telescope. *Kepler* is one of the priority missions of the *Discovery* program. The aim is that of keeping under observation 100,000 solar type stars for four years, with a telescope with a diameter of approximately 1 m. The promoters hope to discover 600 terrestrial planets in this way.

Another proposal for a space mission based on the same principle is the

French *COROT*. Actually, owing to the small size of the telescope, the primary goal of *COROT* is not the search for extraterrestrial planets but astrophysical studies of the stars; it is able to record occultations, and therefore the proponents say it will be able to identify a certain number of terrestrial planets during its observations.

A further European proposal is the *Gaia* mission, whose purpose is to perform accurate astrometric and photometric studies on more than a billion stars. A precise map of the zone of the Milky Way close to us will be obtained, improving our understanding of the formation, composition and evolution of our galaxy. Even if it won't have the direct purpose of looking for habitable or inhabited planets, it will supply extremely useful information in this field.

It is doubtful that observations from the Earth or from circumterrestrial space will achieve more detailed information; terrestrial planets can be discovered and information on their potential habitability, or clues that they are actually inhabited, may be obtained in the future, but to know the forms of life that may exist there, the biochemistry on which they are based, and other such details, we will have to wait until we are able to launch interstellar probes that can study these life-forms in situ and transmit the results to Earth.

There are those, however, who insist that interstellar travel with automatic probes or with crewed spaceships is impossible for any human civilizations; if this were true, the hopes of one day knowing the details of life that developed on extrasolar planets would be in vain. In this sense the supporters of SETI (Search for ExtraTerrestrial Intelligence) are right to think that we will only know about the existence of intelligent life-forms and learn their details if they communicate to us by radio the information that interests us (or, to be more precise, the information they want us to know).

Chapter 4 discusses the possibility of building interstellar probes and exploring nearby planetary systems. Yet it can be safely stated that (a) interstellar travel, though extremely difficult and currently out of our reach, may be possible for civilizations more technologically advanced than ours and (b) the search for extraterrestrial life beyond the Solar System will continue by dispatching probes or spaceships with a crew as soon as technology and economic conditions will allow it. The studies mentioned above, carried out from Earth and from circumterrestrial space, will be useful to perform an initial screening and to choose the objectives, but the true scientific work must be done on the spot.

4

THE SEARCH FOR EXTRATERRESTRIAL INTELLIGENCE

INTELLIGENCE AND CONSCIOUSNESS

IN its search for extraterrestrial life, astrobiology is limited by the enormous distances that separate the stars of our galaxy. As one may assume that interstellar probes will remain outside our possibilities for a long time, the only forms of extraterrestrial life we can discover outside our Solar System are those capable of being detected from a very large distance.

Our present civilization could be discovered from a very large distance thanks to the radio waves we broadcast toward space, so it is a common opinion that only intelligent life-forms can be discovered, either because their technological activities are conspicuous or because they purposely try to contact other civilizations. The various projects collectively known under the acronym SETI (Search for ExtraTerrestrial Intelligence) are based on this idea. In this context, intelligent life-forms are those that either have developed a technology that allows us to discover them, or try to establish interstellar contact themselves. This is a very good example of our restrictive and anthropomorphic definition of intelligence.

Though we understand full well what we mean by the term *intelligence*, we are unable to give a general definition either for human intelligence or for the artificial intelligence with which we try to endow machines. The only definition of artificial intelligence is that based on the Turing test, which states that a machine is intelligent if its behavior cannot be distinguished from that of a human. This definition is an implicit acknowledgment that the only intelligence we know is our own, but is completely useless as a general definition of intelligence.

Human beings are both intelligent *and* self-conscious but, if it may be easier to give a theoretical definition of consciousness than of intelligence, it is much more difficult to tell whether a being is self-conscious or not. Besides, it is not even clear whether consciousness is a "discrete" characteristic (i.e., a characteristic that either is present or is not), or a "continuous" one (i.e., one that may exist in different degrees). Ancient traditions and those of primitive peoples often attributed a true consciousness to animals (at least, to many of them) and even to inanimate beings, in an anthropomorphic view of the world. Until a few years ago, modern science has on the contrary always assumed that the only conscious species is humans and that animals are like automatons. The expression *animal mind* was considered an oxymoron. This idea is also common in the Christian view of the world: human beings, created in God's image, are the only beings endowed with an immortal soul. Consciousness and intelligence are in some way linked with the presence of the soul and are therefore human prerogatives.

The study of animal behavior has recently suggested that some degree of self-awareness may also be present in animals, though this is still very controversial, and now the expression *animal mind* is used sometimes. Undoubtedly the behavior of animals often suggests that they are, at least up to a certain point, conscious, and many owners of dogs and cats are of this opinion. Yet experiments restrict the possibility of a certain measurable awareness to apes and, among them, only chimpanzees have produced some experimental evidence. Besides, if awareness of death must be one of the parameter for self-consciousness, it seems to have been ascertained that even chimps, who show amazement and strong feelings when facing the death of a relative, are not able to extrapolate their own destiny as individuals from it.

So it would seem that self-awareness *is* something that may exist in various degrees, though we can safely state that only the primates closest to humans show some consciousness and that even chimpanzees show just a trace of it. On Earth, consciousness and intelligence appear to be strictly related to each other. Will it be so when (and if) we discover other intelligent beings? Perhaps the only solution, at least for now, is to assume that all intelligent beings are conscious, and that some form of consciousness may be present even in beings that are not yet intelligent. In the field of artificial intelligence, for instance, it has been posited that an intelligent computer (if such a thing will ever be built), able to pass the Turing test and thus be indistinguishable from a human being, must be considered as conscious. Taking the case to its extreme consequences, such a computer should be considered human, and to switch it off would be an act comparable to murder.

As mentioned earlier, one of the prerogatives of life is its ability to store information. On our planet, the information needed to define a living being is stored in the DNA, and each cell of a complex organism contains all the necessary information. It is passed to the new cells when they are produced and from each individual living being to its offspring. When living beings became sufficiently complex they started storing information in another way, namely through memory. We do not yet know in detail quite how this is performed, but it seems to be based on chemical processes, and memory seems to be distributed in the whole body, particularly in its simplest forms. Although even simple beings have some memory, it is with encephalization and the birth of a complex nervous system that it became truly important. Information stored in memory is not passed on to the offspring and is lost when the individual dies.

With greater intelligence, the ability to store information grew rapidly. Through language it became possible to communicate the contents of an individual's memory to another and so give way to a sort of shared memory, belonging to groups of individuals. Then humans invented ways not just to communicate information, but to store information in a more or less permanent way outside the body. This process probably started even before writing (pictograms, ideograms, alphabetic writing) was invented, but with writing it became much more detailed and efficient.

Although nonintelligent beings can also store some information outside their bodies (like the marking of territory by many animal species), we can safely say that this is a typical feature of intelligence and is what allows humans to build a culture, or for different cultures to be built by different human groups. With humans, technology becomes involved in the information storage process: from clay tablets to papyrus, from vellum to printing. Modern information technology radically increased the

Intelligent autonomous and conscious systems

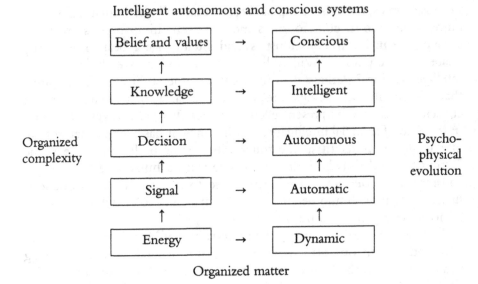

FIGURE 4.1 *Tentative scheme of the path leading from inert matter to intelligent autonomous and conscious systems.*

availability and accessibility of information and it is likely that this trend will continue in the future.

A tentative scheme of the path leading from inert matter to intelligent and conscious systems is shown in Figure 4.1. The diagram has a strong relevance for efforts to build intelligent machines, but it is also applicable to living systems. Note that it is debatable whether the "Decision" box should be over or under the "Knowledge" box. Here it is assumed that a being (living or robotic) can make decisions reacting to the inputs from the outer world even without building an internal model of it: this point has been very controversial,[1] but here it is assumed that a positive answer to this problem is realistic. Moreover, sometimes it seems that there is no complete agreement on the meaning of the terms in the boxes, so that the answer depends on the exact interpretations of *knowledge* and *decision*.

Note that the boxes on the right should not be considered as separate steps in a ladder, but rather as levels in a continuous evolutionary process, and that there are an infinity of shades between each of them.

When we say we are looking for extraterrestrial intelligence, are we looking for only intelligent beings, or conscious beings, or beings like ourselves who are both? When we search for extraterrestrial intelligence,

---

[1] R.A. Brooks, *Flesh and Machines*, Pantheon Books, New York, 2002.

are we actually searching for extraterrestrial minds? Generally, this aspect of SETI is seldom debated, if at all.

CONSCIOUSNESS

While it is self-evident that we are conscious, the origin of consciousness is still unknown. The simplest approach is to state a deep discontinuity between the material world and the world of our mind. In this view, the interiority of consciousness cannot have evolved from the machinery of our body through the usual mechanisms of natural selection. Consciousness is something from outside, from a not-better-identified metaphysical realm. As already stated, this is generally (but not always) the approach to consciousness of the various religions. It must, however, be remembered that well-known scientists also agreed with this approach; the most famous is Alfred Russel Wallace, the co-discoverer with Darwin of the theory of evolution, who thought that human consciousness "could not possibly have developed by means of the same laws which have determined the progressive development of the organic world in general, and also of man's physical organism."[2] He thought that some metaphysical force had directed evolution at three points: the beginning of life, the beginning of consciousness and the beginning of culture.

Actually the problem of the soul, strictly linked with consciousness, is one of the three problems (the soul, the Cosmos, and God) that Kant thought we cannot solve by theoretical reasoning (*Critique of Pure Reason*). However, science has the precise task of explaining all it in terms of natural phenomena only, without resorting to metaphysical entities. Many phenomena that in the past were explained resorting to metaphysics were eventually explained purely in terms of the physical world, so it is possible that in the future what Wallace thought impossible may be achieved.

Note that if consciousness is something introduced into the world from outside, the last step of Figure 4.1 is something qualitatively different from the others. The possibility that conscious beings developed in other places in the Universe would in this case only depend on specific metaphysical interventions taking place somewhere else. To avoid metaphysical interventions on the physical world, a first possibility is to assume that consciousness is an intrinsic property of matter. But, notwithstanding the

---

[2] A.R. Wallace, *Darwinism, an Exposition of the Theory of Natural Selection*, Macmillan, London, 1889, p. 475.

efforts of its proponents, this seems a far-fetched hypothesis, evidently influenced by our generic tendency to attribute our own character to everything around us.

Along this way, an interesting solution to the problem of how intelligence and consciousness developed is the so-called complexity–consciousness law, formulated by Teilhard de Chardin: consciousness, in its basic form, is intrinsic to matter as such, but it emerges only with increasing complexity in the form of living matter. In this sense, evolution would be a continuous trend toward complexity and therefore toward consciousness and intelligence. With human beings, matter becomes fully conscious and starts the formation of a noosphere, the sphere of conscious thought, that adds itself to the biosphere, just like the latter enveloped the lithosphere of the planet (or, according to recent discoveries, at least partially penetrated it).

This interpretation is often considered too finalistic and is opposed by those who think that evolution does not proceed in any particular direction but at random, and that the increase in complexity is just a fortuitous aspect of diversification. Clearly if the complexity–consciousness law has any validity, intelligence is, if not a necessary result of evolution, at least a very likely outcome, whereas if evolution proceeds completely at random it may be a casual, and perhaps extremely rare, event.

Another possibility is that consciousness is a property of protoplasm, that is, of living matter. Darwin himself was of this idea, like many other nineteenth- and twentieth-century scientists. The behavior of even the simplest organism seemed to them to be dictated by consciousness, a statement strongly vitiated by anthropomorphism. The untenability of this position pushed other scientists to assert that consciousness developed later, together with the ability of learning from experience, and is strictly linked with it. A good confutation of this theory can be found in *The Origin of Consciousness in the Breakdown of the Bicameral Mind*, by Julian Jaynes.[3]

The following theory is defined by Jaynes as the *Helpless Spectator Theory*. Consciousness is a consequence of the trend toward a greater complexity of the nervous system and appears as a stage of the evolutionary process; however, it has nothing to do with our behavior. How we behave is just the result of the "wiring" of our brain and consciousness is just a sort of optional add-on with no influence on what we do. To put it as Thomas Henry Huxley—Darwin's bulldog, as he was nicknamed—did, "we are conscious automata." If this were true, consciousness would be fully immaterial in this

---

[3] J. Jaynes, *The Origin of Consciousness in the Breakdown of the Bicameral Mind*, Houghton, Boston, 1976. The author is deeply indebted to this book for the ensuing parts of the present section.

context, and we should deal with it no more. By the way, it will also be of no importance in any other context! Needless to say, this theory was strongly refuted by many scientists—and not only by them.

To restore some role to consciousness, another explanation was given: consciousness is an emerging property of evolution. The concept of emerging properties was put forward in the 1920s in this context, but has recently been used to explain the behavior of complex systems and has become very fashionable in conjunction with chaos theory. Essentially, it is based on the idea that the properties of a complex system cannot be derived completely from those of its constituents: the reductionist approach misses something when applied to complex systems that are not just the sum of their constituents. Just as the properties of living organisms cannot be derived from the properties of the molecules of which they are made, living beings, when they become complex enough, acquire properties that cannot be understood from those of their biological machinery. The problem with this approach is that it does not explain when, along the evolutionary path, consciousness emerged, and certainly does not explain how it works. Also, in a way, it is not very far from the complexity–consciousness theory outlined above, at least in its consequences.

As a reaction to the generalities of the emerging properties theory, the *Helpless Spectator Theory* was revived in an even more extreme form: behaviorism. Its basic idea is that consciousness not only does not have a role, but in fact does not even exist. In its modern form it started after World War I and dominated psychology from 1920 to 1960. It reduced everything to a series of reflexes to external stimuli, and had its roots in a myriad of experiments on animal behavior performed in laboratories in the whole world. As far as our context is concerned, a behaviorist approach to SETI would be to just search for intelligent signals without bothering to ask ourselves whether their authors are conscious or not, since that question has no meaning. All extraterrestrial intelligences would be just intelligent automata, like ourselves and the intelligent machines we are going to build. It also does not make much sense to wonder whether a possible contact is with living beings or robots: they are not qualitatively different, and it matters little whether the beings we get in contact with are made of carbon, other types of biological material, silicon and steel, or whatever.

But even if we say that consciousness does not exist, in that moment we are conscious of doing so! We could accept, perhaps with difficulty, that our consciousness has no influence on our life, but that it does not exist goes against our perception of reality. The problem is back; we feel we are conscious, but where does consciousness come from? Instinctively we feel it is linked with our nervous system, and in particular with the brain, but where in the brain can it be located? Descartes thought it was located in the pineal body of our brain, but this idea was refuted. Much research aimed at finding the seat of consciousness went on, and is still active. A solution advanced by many is the reticular formation, or reticular activating system as it is called with reference to its function. This is a network of neurons, connecting the whole brain, that activates and deactivates the various nervous circuits. However, this is one of the most ancient parts of the nervous system and there is no serious clue to indicate that it is connected with consciousness.

A metaphor of the working of our mind states that it is like a computer, with our brain playing the part of the hardware, with a software that gives us our intelligence and consciousness. There is nothing new in this; in all ages there were metaphors describing the working of our body with reference to the most modern (for the times) machinery. We were described in terms of clockwork mechanisms (with God as the ultimate clockmaker), steam engines, and so on. There is no problem when these are metaphors, but now the similitude with the computer is intended by some in a literal way. Frank Tipler takes the analogy so seriously that he says our brain is actually a biological computer and our mind is a creation of the software running on it, a software developed step by step by evolution as the hardware became increasingly complex. He takes for granted that it will one day be possible to transfer our mind onto another hardware, perhaps a very advanced digital computer, in the same way that any program can be moved from one computer to another. This is taken for granted, like a corollary of the postulates of hard artificial intelligence (AI). However, no proof is produced either for AI or for this view of human consciousness.

Finally, let us consider the importance of consciousness in our intelligent behavior. As already stated, some hold that there is no such thing, and that we are just automata. But if this is an unlikely statement, the opposite one is also unlikely to be true. By definition, we do not realize how much of our behavior is carried out without our awareness, but in many instances we receive the results of some mental processes without even being conscious that we were working on those issues. Examples in the artistic, scientific, and everyday life fields are too common to be mentioned here. Artistic creation often (even usually) surfaces to our consciousness in a complete state, as if coming from outside our self. Many activities requiring fast reactions (driving a vehicle, playing sports, operating a manual machine tool, etc.) are done in a seemingly automatic way and could not be otherwise, since conscious reasoning would be too slow. For low-level reactions this is usually ascribed to training, but often higher level actions are also done without "consulting" our conscious self. In driving, for instance, not only obstacle avoidance, gear shifting, and other rapid reactions are done by the automaton in ourselves, but also navigational tasks. When driving on a well-known road we let our inner automaton choose the way, just as we would let a horse we are riding follow the way it knows.

But if we are not intelligent automata, is it possible that some extraterrestrial intelligences are? This point will be discussed again when dealing with human evolution to the conscious stage. The only positive facts are that intelligence is strictly linked to the brain and therefore its appearance is linked with the development of the nervous system and the brain and that the only intelligent being we know is also conscious. There is, however, the possibility that intelligent and conscious beings based on completely different systems may exist. On Earth the development of the nervous system has brought to the concentration of nervous functions in a single organ, the brain: neurogenesis led to encephalization. Is this just a casual circumstance? Could intelligence based on a completely different architecture exist, for instance a nervous system that is highly distributed in the whole body, or does any intelligent being necessarily have a big brain, possibly placed inside a protective box?

THE DEVELOPMENT OF INTELLIGENCE ON EARTH

When eukaryotes started to associate into something more organized than simple colonies of cells, they started to differentiate. The first, very simple, multicellular organisms were born. In the animal kingdom the first real metazoans, the coelenterates, began to have specialized cells performing control functions—nervous cells. Neurogenesis is strictly linked with multicellularity. The complexification and differentiation of life, as shown in Chapter 3, underwent an impressive acceleration during the Cambrian period, about 650 million years ago. In this period all the configurations of living organisms that still exist (plus many that are now extinct) appeared and neurogenesis continued along the two main roads of complexification of the nervous system and its encephalization, that is, the concentration of the control functions in a single organ, the brain, of continuously increasing size.

The nervous system of the invertebrates has the form of a chain of ganglia extending along the whole body, but in this case, too, a progressive fusion occurred in the ganglia located in the front part of the body, to form a cerebral ganglion placed close to the eyes. Starting with the first chordates, the nervous system organized itself into a single structure, the notochord, with a cerebral vesicle containing the nervous cells in its forward part. In the vertebrates the latter becomes a real brain, surrounded and protected by a box of bones produced by the forward part of the skeleton. With the exit of animal life from the sea and the development of amphibians, reptiles, birds, and finally mammals, the brain continues to increase its complexity and differentiate internally. The cerebral cortex becomes more and more important, until it becomes the center that controls voluntary movements in mammals.

After the K/T extinction, mammals obtained a sort of supremacy, at least as far as land animals of a large size were concerned; it must be remembered that if we measure the success of a type of animal by the number of species or individuals, then bacteria are the most successful organisms that ever appeared and, among the animals, insects are much more successful than vertebrates. The first primates date back to the cretaceous period, more than 100 million years ago, before the K/T extinction. In the beginning they were mainly arboreal animals, with four limbs of five fingers each and a plantigrade posture. The characteristic short snout allowed the eyes to be in a frontal position, which in turn afforded binocular vision and, perhaps more importantly, put the front legs

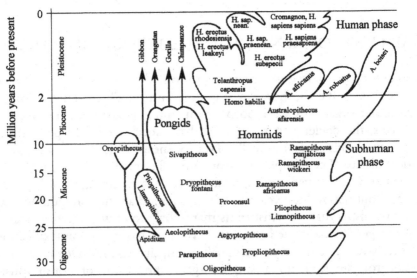

FIGURE 4.2 The Hominoidea. Evolution of humans and apes starting from common ancestors in the Oligocene, about 30 million years ago. Note that the time scale is strongly nonlinear (it is greatly expanded toward the present).

inside the animal's field of vision. After the disappearance of the dinosaurs, at the beginning of the Paleocene (see Figure 3.9) primates were similar to the present tarsiers, and between 50 and 40 million years ago the order differentiated into two suborders, the lemurs and the monkeys—the latter having already divided into the families of the platyrrhine and catarrhine monkeys.

At the beginning of the oligocene (37 million years ago) the hominoid branch detached from the monkeys (Figure 4.2); from that detachment apes and humans would evolve. After the separation of the gibbons, the differentiation between the pongids (orangutan, gorilla, and chimpanzee) and the hominids occurred about 10 million years ago, probably with the birth of the Australopithecus. This evolutionary event occurred in East Africa and marked the appearance of a species with larger brains (Figure 4.3), a size never reached before, at least in terms of the ratio between the weight of the brain and that of the body.

The fact that humans evolved from primates, and that we are close relatives of apes, is one of the most important, and at the same time initially more controversial, results of the theory of evolution. The basic idea was already clear in the middle of the nineteenth century, but the details are still not completely clear today. Darwin thought, against the opinion of the majority of scientists and the fossil evidence available in his time, that the cradle of humanity was Africa. His idea was that several million years ago, perhaps more than 10 million, some primates began to walk with an erect posture and that this allowed the hands to adapt to perform more delicate jobs than simply bearing the weight of the body: manufacturing objects and throwing well-aimed weapons. These bipeds, handling tools and weapons, developed more intense social relations, something requiring greater intelligence. This scenario was not just a scientific statement: the differentiation of humans from apes was both ancient and sudden and this set a large gap between us and the apes.

But fanciful ideas on the origin of the human species were still common in the beginning of the twentieth century. In 1910, for instance, Hermann Klaatsch suggested a genealogical tree for humans according to which the apes had separated into four branches, from the first of which Tasmanians and Australians originated, from the second gibbons and the Asian races, from the third chimpanzees and the black race, and from the fourth orangutans and Europeans. The possible racist implications of theories in which the various human races originated from different ancestors is obvious.

Only with the progress in the field of genetics and the discovery of DNA did it become clear that the differences between human races are very small and that the apes genetically closest to humans are the

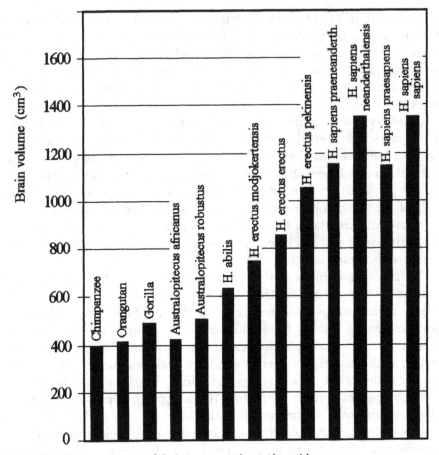

FIGURE 4.3 Average volume of the brain in apes, hominids, and humans.

chimpanzees. The two species have almost all chromosomes in common with the exception of five that have a pericentric inversion and two chimp chromosomes that are fused into a single one in humans. The first, therefore, has 24 pairs of chromosome, while the second has 23.

Darwin's explanation, according to which the biped stance, the production of tools and the need of greater intelligence for social interactions would have been developed together, no longer holds today. The fossil evidence and molecular biology studies allow us to state that around seven million years ago, in the Rift Valley of East Africa, a species of apes evolved that were able to walk on two legs. Their behavior was typical of apes and the volume of their brain was little different from that of other apes. Differentiation probably started 12 million years ago, when large tectonic upheavals caused the Rift Valley to sink, dividing East Africa into two parts and causing strong climatic changes. The apes living west of the Rift Valley, in a humid and wooded environment, gave rise to modern apes, while those that remained in the zone east of it, drier and covered by bushes, originated the hominids.

Anthropologists found the fossilized remains of biped apes of different species, which were given the name of Australopithecus (the famous skeleton found in 1974 and denominated Lucy was a young female of this species). As always, when evolution accelerates and nature seems to experiment with new forms to fill an ecological niche, various species bloom, correlated with one another. Later, natural selection will perform its task of choosing which of them will survive and which will become extinct.

In this blossoming of species within the hominid family, hominids quite different from Australopithecines appeared around 2.5 million years ago; they had a less stout build, smaller teeth, and a larger brain. The remains of this new species are normally found together with stone tools, and for this reason it has been given the name of *Homo abilis*.

A relatively short time later, another new species appeared, which was named *Homo erectus*.[4] The new hominids had an even larger brain and continued to evolve until the volume of their brain was more than 1000 cm^3. Their appearance was accompanied by stone tools of a new, improved type. *Homo abilis* and then *Homo erectus* shared their habitat with the Australopithecus for more than a million years, up to the time the latter disappeared. The role of *Homo erectus* in the extinction of the Australopithecus has long been debated but, even if it is likely that the first may have used the second as food, the extinction of the Australopithecus was more probably the result of competition than of direct hunting.

One of the crucial steps on the way leading to intelligence seems to have been the evolution of the *homo* species from the Australopithecines; the latter were just biped apes that maintained many of the typical characteristics, and probably the behavior, of apes, while the homo family were something different from the beginning.

A beautiful reconstruction of the behavior of *Homo erectus* about 1.5 million years ago can be found in Richard Leakey's book, *The Origin of Humankind*.[5] It describes a day in the life of a band of *Homo erectus* at "Site 50," an archaeological site close to the Kazari Escarpment, about 20 km

---

[4] No distinction is made here for simplicity between *Homo erectus* and *Homo ergaster*. The latter, who preceded the former by about 100,000 years, might have been a separate species or a subspecies of the first one. Its appearance marked a sharp progress in stone tools.

[5] R. Leakey, *The Origin of Humankind*, Orion Books, London, 2000, pp. 93–98.

east of Lake Turkana, in northern Kenya, where abundant traces left by *Homo erectus* have been found. Their behavior is described in very human terms: these hominids look to all effects human. If astronauts from Earth, landing on a distant planet, should find a group of this kind, regardless of the form of their body or the biochemistry on which their life is based, they would recognize intelligent beings, endowed with at least some form of self-consciousness. They should be considered humans, with all the consequences that entails, beginning with recognizing in them all the rights human beings have.

Not all researchers would agree with this picture; some think that 1.5 million years ago *Homo erectus* was much more similar to apes, but the ideas on which Leakey's reconstruction is based are now more widespread. The study of the fossils and stone implements discovered at various sites and the progress in the field of genetics suggest that anatomically and from the behavioral point of view the true passage from apes to humans occurred when *Homo abilis* evolved from Australopithecus. However, this does not mean that these humans were conscious in the sense we apply to ourselves.

From the anatomical point of view, even if no fossil traces of the brain may remain, its form leaves an imprint on the inside of the skull and therefore a casting of the inside of the braincase yields a model of the outside of the brain. Operating in this way, it is possible to see that the cerebral structure of the Australopithecus is essentially ape-like, while that of *Homo erectus* is human.

The reconstruction also shows the importance of eating meat for early humans. The brain consumes large amounts of energy: it constitutes only 2 percent of the body weight, but consumes 20 percent of the available energy. To be able to maintain a brain of a large size, humans need food that can be assimilated quickly and supplies much energy. The need to coordinate individual efforts for hunting—in the absence of a body specialized for this way of obtaining food (humans do not run particularly fast and are not endowed with fangs, claws, or other natural weapons)—and the increase of time freed from the primary function of feeding, surely had a positive effect on the development of intelligence.

The birth of intelligence and a carnivorous diet seem therefore strictly linked to each other. Australopithecus was probably partially carnivorous, but his less athletic body and his uncertain bipedal walk due to a still ape-like anatomy (his organs of equilibrium in the inner ear were much less evolved than those of the early *homo*) made it more of a scavenger—looking for animals killed by other predators—than a hunter. Besides, the absence of teeth suitable for a carnivorous diet made it really difficult to feed on meat without the help of tools.

Still, what is more impressive in the hominids of Leakey's reconstruction is their loquacity. But had *Homo erectus* developed real human language 1.5 to 2 million years ago? Scientists were once sure he had not; language was seen as a very recent conquest in human history. Today they realize that this conquest was probably gradual, like the anatomical modifications that made it possible. The conformation of the larynx of the Australopithecus is surely ape-like, while to see a really human larynx it is necessary to wait until much later, when *Homo sapiens* appeared. *Homo erectus* had an intermediate configuration, which probably allowed him to articulate a range of sounds well beyond what is possible for animals. Another sign is the slightly larger development of the left part of the brain, where the centers related to speaking are located. This characteristic, present in the majority of humans, is also linked to the fact that, in general, humans use the right hand better than the left one (the right part of the body is controlled by the left part of the brain). The skulls of *Homo erectus* seem to show this asymmetry and the shape of the stone tools he made suggest that the job was performed mainly using the right hand.

It therefore seems possible to state that between one and two million years ago, *Homo erectus* was able to communicate with his fellows in more articulated and complex ways than apes do at present and Australopithecines ever did. But was he able to speak a true language? The sounds and gestures animals use for communicating are typical of each species, while human oral language, and in part the language of gestures, too, is a product of culture and changes according to place and time.

The success of *Homo erectus* is testified by the fact that other hominids of that species were found well outside his East African zone of origin, east of the Rift Valley. The specimen with the largest brain was found in China: the brain of the *Homo erectus pekinensis* was larger than 1000 cm^3. The ability of early humans to thrive in ecosystems so different from those in which the species evolved is important. With intelligence, a "generalist" animal appeared who could adapt to a wide variety of environments and lifestyles, instead of many "specialist" animals, each one suited to a certain lifestyle and a certain environment. As will be seen in the next section, technology played an essential role in the ability of *Homo erectus* to adapt to various environments and become the dominant living being in each of them.

Slowly the habitat of humans became larger and in a few hundred thousand years it extended to the whole of Africa and Eurasia, or at least to the zones of these continents that were not too inhospitable. This is not amazing in itself: if a community moves its campsite by 10 meters every year, always in the same direction, after 100,000 years it will dwell at a distance of 1000 km from the point of origin.

Homo erectus was well on the road leading to intelligence: he built simple tools and weapons of stone and was able to communicate in some way with his fellows, even if his language abilities were probably very limited. Was he also self-conscious? Obviously it is not possible to give a sure answer, but it is very probable that the consciousness of these early humans was very limited.

About two million years after the time the first *Homo* (first *abilis*, then *erectus*) appeared, a new evolutionary step was climbed: a new increase in brain volume and other anatomical changes, accompanied by a true technological revolution in the method of producing stone implements. A new player appeared on the stage of history: *Homo sapiens*. The new species expanded fairly quickly in the whole habitat previously occupied by *Homo erectus* and, while initially living together with the latter, finally replaced him. It has long been discussed whether this evolutionary step occurred again in East Africa and a new migratory wave started from there or if it occurred separately in the various populations and gradually changed *Homo* from *erectus* to *sapiens*, through the importation of genes in cross-breeding with outsiders. Today the first theory seems more likely.

The new human species that appeared about half a million years ago represented large progress in comparison to the previous ones, but was not yet fully modern, in the sense that those humans were not yet really similar to us. Slightly more than 100,000 years ago a variety of *Homo sapiens* appeared in Europe and part of Asia, probably more suited to the intense cold that was prevalent there because of the glaciations: the Neanderthal man (*Homo sapiens neanderthalensis*). It is not clear whether it was genetically a completely separate species or just a variety, and it is uncertain whether the two branches of *Homo sapiens* really lived together in the same territory at the same time or if waves of populations of the two types inhabited much of Europe one after the other, depending on the climate.

Around the same time Neanderthal man appeared, modern man, *Homo sapiens sapiens*, also made his first appearance. Here, too, there are conflicting theories: modern humans could have developed from pre-modern ones in the various places, or they could have migrated from a well-determined place of origin to substitute the previous populations. Studies of the DNA present in the mitochondria brought to the formulation of the so-called *mitochondrial Eve* hypothesis: all modern humans could be the descendant of a single woman living 150,000 years ago, perhaps in Africa. Naturally this does not literally mean that we all descend from a single couple, an African Adam and Eve, but that a woman belonging to a population of at least 10,000 people, with all the interbreeding that this makes possible, was the earliest progenitor to have the same genetic information in her mitochondria that we carry today.

The brain volume of the new species, as well as that of the Neanderthal man, was the same as that of modern humans: an average of 1350 cm^3. For the first time, slightly more than 100,000 years ago, we find the trace of a real human mind at work: the deliberate burial of a dead body. To this first burial many others followed, with the body always arranged in the same positions within each culture, with objects and often traces of ritual acts. Clearly the absence of tombs does not mean the absence of consciousness, or of a view of the world in which life and death have a meaning, but it is a strong clue for it.

The presence of burial sites is a strong indication that the two human species that shared supremacy on our planet (although Neanderthals were actually present only in central Europe and in the Near East up to the Caucasus) in the last 100,000 years had a consciousness similar to ours and were human to all effects. Starting 40,000 years ago we have other indications of human consciousness: the appearance of real art works, rock paintings and sculptures that we still can admire, and probably paintings on perishable material like hide, which left no traces. Such archaeological evidence, and above all the beautiful rock paintings (Figure 4.4) found in many caverns, testify that humans reached a sensibility and a capacity for abstraction that cannot be explained if we do not admit that their intelligence was very similar to ours. Many went further and asserted that a *sapiens sapiens Cro-magnon man* would be physically and intellectually indistinguishable from a modern man.

Notwithstanding what has been said above, some scientists believe that consciousness is a much more recent acquisition. Jaynes, in *The Origin of Consciousness in the Breakdown of the Bicameral Mind*,[6] advances the theory that human consciousness developed less than 3000 years ago; all humans living before that time were just automata, intelligent ones perhaps, but unaware of themselves. If a civilization could actually reach that stage, which included both technology and art (from the Iliad to the Pyramids, from Gilgamesh to the Temple in Jerusalem) without consciousness, it could undoubtedly also build radiotelescopes and other

---

[6] J. Jaynes, *The Origin of Consciousness in the Breakdown of the Bicameral Mind*, Houghton, Boston, 1976.

FIGURE 4.4 *Rock painting of a bison, dating back to about 17,000 years ago. Cave of Altamira (Spain).*

means to contact their cosmic neighbors. However, this hypothesis has very little following.

Only one of the various species of hominids that developed in the last two million years survived, and at a certain point it apparently stopped evolving physically, shifting its progress to the field of cultural evolution.

It is not known how the other species became extinct, and the end of Neanderthal man, who disappeared about 33,000 years ago, is particularly mysterious. The disappearance of this species begins in Asia and then in eastern Europe about 50,000 years ago, and was completed in western Europe about 17,000 years later, following a geographical pattern similar to that of the population by *Homo sapiens sapiens*, but with a delay of about 50,000 years. It is likely that members of the two species came in contact and that they fought against each other. If the Neanderthals were intellectually inferior to the *sapiens* (particularly as far as language is concerned, to judge from the anatomy of the larynx, and in spite of the volume of the brain), the result of such clashes was unquestionable. Between the opposite theories, the elimination of the Neanderthals by the *sapiens* and its assimilation into a single species from which modern humans originated, the first one is probably more realistic.

Even before *Homo sapiens sapiens* became the only intelligent species on the planet, humans spread, probably in various waves, in all the climatically fit regions of Eurasia and Africa. About 55,000 years ago they entered Australia and expansion on all continents was completed around 20,000 years ago, when the Strait of Bering was crossed and they entered America. Their ability to build stone and wood tools and weapons, to prepare at least provisional shelters, to cover themselves with animals skins and to use fire gave them a new advantage in the struggle for survival and allowed their habitat to extend to zones that, without suitable technology, would have been denied them. As noted earlier, this superiority in the struggle for survival caused the extinction of many species, and humans became themselves a cause of mass extinctions. Their behavior also made humans a strong agent of ecological change, in many cases certainly not for the best: man's role may have been determinant, for instance, in the desertification of Australia.

Yet, as long as humans were just hunters and gatherers of spontaneous vegetables, their role in changing the planet was bound to remain rather marginal.

FROM INTELLIGENCE TO TECHNOLOGY

Since the beginning of their history, humans qualified as technological animals. The ability to create technology is probably identical with intelligence, and the presence of an intelligent being can be inferred from archaeological evidence of a technology of some type. In the classification of the human species that appeared one after the other, *Homo habilis* precedes and prepares the ground for *Homo sapiens*. Some animals, for instance, are able to use objects to perform actions, as seals do when they use a pebble to break the shell of a mollusk or certain apes when they use a stick as a club. But while some animals may occasionally use an object to perform an action, only humans, even the most primitive ones, are able to build a tool in view of its use and then keep it to perform future actions.

Actually, when faced with the traces of an extremely primitive technology it is impossible to deduce the actual presence of an intelligent and conscious being in the sense that we currently give these terms: the *Homo erectus* of a million and a half years ago, for instance, was certainly able to build and use simple stone tools, but it is unknown whether he actually possessed intelligence accompanied by consciousness. The same can be said for the use of fire: traces of it can be dated as far back as 700,000 or 800,000 years ago, when human consciousness was very unlikely, as mentioned earlier. After all, if *Homo erectus* could build stone implements, why could he not maintain fire for his own use or even light it?

Sometimes intelligence is said to have little or no value for survival and therefore not to be particularly favored by evolution, but on technology no doubt can exist: its value from this point of view is enormous. The ability to build tools is surely favored by natural selection. One may doubt that a nontechnological intelligence can exist, since it is unrealistic that an intelligent being would not develop a technology that allowed him to solve problems linked to his own survival, but above all because intelligence alone, without technology, would not be useful in the struggle for survival, even if that struggle is not meant in the obvious sense of overpowering potential enemies and getting food, but in the deeper sense of propagating genes and leaving as many descendents as possible. Besides, the necessities of tool construction and those of hunting and a complex social life have often been considered among the causes of the increase in size of the brain and the complexification of its structure.

Technology, particularly in the beginning, consists of the manufacture of objects that increase the potentialities of the human body, objects that work like prostheses: an axe to make up for the lack of claws or fangs, a skin garment to make up for the lack of a fur that gives protection from the cold, and so on. The things that other species accomplish by modifying their body slowly, humans obtain in an incomparably shorter time by creating purposely designed objects.[7] Humanity therefore no longer needs to evolve physically; it can take control of ecological niches that were occupied by different animals, each one with its specialized body. Or better still, it evolves by refining its mental abilities and the individual's interaction with other members of the same species, as shown by the increasing size of the brain and the complexification of the larynx.

The main advantage of adapting to the environment by adding technology to one's physical attributes instead of evolving slowly is the quick reaction to environmental changes, something that has enormous value for survival. However, it must be recognized that this is not free from dangers, since in this way it is possible to cause rapid changes, so quick that the environment of humans may not be able to deal with them with the necessary speed. The extinctions caused by humans are a clear example: animals in Eurasia learned to fear this new competitor as he evolved, and to defend themselves against him. When humans landed in Australia and America with a developed technology, they learned to hunt the new species quickly, far more quickly than animals could learn to stay away from the newcomers. The result was the extinction of many species and this later influenced negatively the development of humans in those

[7] E. Righetto, *La Scimmia Aggiunta*, Paravia, Torino, 2000.

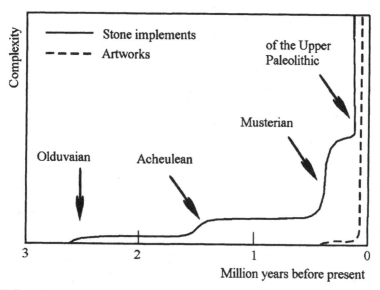

FIGURE 4.5 *Increasing complexity and variety of stone tools and complexity of artworks in the development of humanity. Note the discontinuities linked with the appearance of* Homo abilis *(Olduvaian culture),* Homo erectus *(Acheulean culture), primitive* Homo sapiens *(Musterian culture), and finally of the* Homo sapiens sapiens *and* Homo neanderthaliensis *(culture of the Upper Paleolithic).*

lands, because in the following phase they could not find other species suitable for taming.

Technological development went on in a very discontinuous way. The variety and quality of stone tools show rapid changes followed by periods of stasis of hundreds of thousands of years (Figure 4.5). The periods of change coincide with evolutionary stages that brought into existence new species of *Homo*. It therefore seems that members of a certain species reached the highest level permitted by their brain in a short time (*short* in evolutionary terms, obviously), and then continued to replicate the same technology. Also the fact that the objects produced by a certain species show very limited local variations seems to point to the technological abilities being much less influenced by culture than by physical structure. If a comparison of this type makes sense, it is something like rigid automation, in which automatic machines always produce the same object, defined by their structure (hardware), as opposed to flexible automation, in which the machine produces a variety of objects, according to the program (software) that has been introduced in it.

When *Homo sapiens sapiens* and Neanderthal man appeared, they brought a revolution that gave life to the culture of the upper Paleolithic. Only with this event do we become sure that we are dealing with actual human beings in the full sense of the term, and artistic expression becomes a central fact of culture. The shape of stone tools, but now also of implements made of bone and other materials, is no longer linked only to their function: now they are usually decorated, sometimes in truly valuable ways. And tools and artworks take on forms and styles that depend on the cultural environment: humans differentiate clearly in various cultures, inventing different styles and technologies.

When *Homo sapiens sapiens* had spread in almost all continents and got rid of his last direct competitor, Neanderthal man, a new phase began, usually referred to as the "neolithic revolution." Slowly he learned how to cultivate some vegetable species and raise some of the animal species he was feeding on, instead of gathering spontaneous fruits of the earth and hunting wild animals. This fact had such strong consequences on the future development of humans that we cannot imagine an evolved intelligent species not having passed through this stage. Humankind did not go through the neolithic revolution simultaneously in the whole planet, and results were different in the various geographical areas; it is even possible to trace the different development of the various peoples and recent phenomena like colonialism and underdevelopment back to the differences in the time when this stage was reached. As Jared Diamond notes, these different times are not due to differences in intelligence or nature of the various populations, but to geographical differences among the regions in which they lived and to differences among the animal and vegetable species of the various continents.[8]

If, therefore, the neolithic revolution is a necessary stage in the development of an intelligent species, the geographical conformation of the planet on which it lives and the potential for domestication of its flora and fauna may have large effects on the type of intelligence, perhaps on the very possibility of continuing along the road leading to a real noosphere. For instance, on a planet in which dry lands are separated into many islands by wide seas or in which the continents are mainly oriented in a north–south direction and have a small east–west size, an intelligent species might not succeed in going through the stage leading from a hunter–gatherer economy to an agricultural–pastoral society, without which the advantages of the development of intelligence remain limited. In such cases evolution could simply take other roads.

Is the neolithic revolution a stage all intelligent species must go through in their cultural evolution, or is the very concept of neolithic revolution only applicable to humans on Earth? If by "neolithic revolution" we mean

[8] J. Diamond, *Guns, Germs and Steel*, W.W. Norton, New York, 1997.

the process leading to the ability of producing its own food, without depending on the uncertain chance of finding it in the form of spontaneous plants or game, it seems reasonable that all intelligent species, with the possible exception of autotrophic species, must go through that stage. May an autotrophic species be intelligent? Obviously we do not know, but if the development of intelligence is at least partly linked to the need to find food, a reasonable hypothesis is that the ease with which an autotrophic species solves its food problems makes this hypothesis at least unlikely.

Actually the neolithic revolution is much more than the ability to produce food instead of gathering it or hunting it; by domesticating animal and vegetable species, humans gradually modify them with respect to the wild ones from which they descend. Domesticated species acquire characteristics that are of advantage not to themselves but to the domesticating species. A sort of protected environment is thus created, dominated by artificial selection, in which the usual rules of natural selection do not apply. Besides, domestic animals not only supply proteins but also perform other functions that are essential for the development of a technological civilization, from supplying energy to producing fertilizers.

If every intelligent species, in its development, must go through a neolithic revolution, can this stage be taken for granted or could there be cases of intelligent species that remained locked in the previous stage, without reaching a true technological civilization? Once the neolithic revolution started, humankind entered a phase of quick technological progress. The progress had an impressive acceleration from the beginning: paleolithic cultures lasted tens of millennia, during which no apparent changes of the material base of life occurred. For long periods humans always flaked stones in the same way, always obtaining similar implements. From the time they started cultivating, the characteristic times shortened, first becoming of some millennia and then, after humans discovered how to work metals, of a few centuries. With the accumulation of technological know-how and the possibility of distracting an increasing number of people from activities directly connected with the production of food, the characteristic times still shortened. Today the material base of life changes within decades.

This acceleration of progress necessarily has a limit, because it strains human adaptability; the refusal of technology, withdrawing into a "natural" world that only exists in the imagination, so common nowadays, is a symptom of it. For many people all technologies that are old enough to have been, so to say, metabolized, are perceived as "natural," while new ones are perceived as artificial. Such a distinction has no objective value, but is a symptom of an uneasiness caused by the acceleration of technological progress. The speed of change makes it impossible for us to imagine what human life may be like just 50 or 100 years hence, just as 100 years ago it would have been absolutely impossible to imagine the modern world. All this gives an idea of frailty: if changes are so quick, it seems that the duration of an intelligent technological species must be quite short. The observation that technology gives humankind more and more powerful means of mass destruction strengthens this feeling. If this were correct, the number of intelligent species in our galaxy at any given time would have to be extremely small.

But it is not just a problem of self-destruction; what really seems impossible is progress that continues at the present pace for a very long time (in the order of millions or tens of millions of years). On the other hand, this is the time scale humans must face when reasoning in cosmic terms. As will be seen below, if a technological civilization is not able to last for such long times, the possibility of the simultaneous existence of several civilizations at distances allowing contact between them is negligible.

But is it possible for a technological civilization to slow its rate of development, once it has reached a certain stage? We do not lack historical examples, even if in different situations: Jean Heidmann reports the case of *Homo abilis* who, having reached a stage of maximum development, apparently remained locked in this cultural stage for more than a million years (Figure 4.5).[9] The situation only changed with a biological development that caused a new species to emerge. In the case of modern humans it is difficult to imagine a situation of this type, but it cannot be ruled out that a stabilization, this time at a different level, could occur again.

The idea that scientific advancement has limits, and that we may even be quite close to them, has often been advanced[10] and even substantiated by alleged evidence. Following this line, it has been suggested that our present phase of quick technological advance will end within a couple of centuries, so that civilizations much older than ours are only about 200 years ahead of us as far as technology is concerned.[11] Such reasoning is not, however, very convincing. We know that scientific and technological progress does not move at a constant pace, and history shows that periods of rapid advances are followed by periods of stagnation, but this does not

[9] Jean Heidmann, *Extraterrestrials*, Cambridge University Press, Cambridge, 1996.
[10] J. Horgan, *The End of Science*, Helix Books, New York, 1989.
[11] P. Musso, "How Advanced is ET?," *7th Trieste Conference on Chemical Evolution and the Origin of Life: Life in the Universe*, Trieste, September 2003.

mean that an alleged present slowing down will lead to a terminal petering out of all progress. It is possible that there are limits to technological progress and if the human species starts diffusing in space, first in the Solar System and then in other systems, the enormous distances between the centers of population and the speed limits that, according to modern physics, cannot be overcome, will cause a slowing down of the pace of development and, probably, a stabilization.

Besides, the human life span is continuously increasing and it is predictable that in the future humans will live much longer than today. It seems that the process of aging is controllable, at least up to a certain point, though there is no clear idea of how much longer human life may grow. Some even think that in the distant future a sort of immortality may be reached, at least in the sense of an indefinite extension of human life. The hope that scientific progress will lead to this result is ancient and the search for immortality is perhaps as old as humankind. At any rate it could be only a relative immortality, since accidental death cannot be eliminated. More realistically, it is reasonable to expect that the average life may become longer than that of the oldest people now alive, and a near-term goal may be a life span of one century. An extension of the duration of life must be accompanied by a decrease of the birth rate, and this will slow down the pace at which generations come to positions of responsibility. A predictable consequence of this mechanism is a slowing down of innovation in all fields, particularly the acceptance of true scientific revolutions that usually requires a new generation of scientists to substitute for the older one.

If these scenarios of stabilization and deceleration of technological progress are reasonable, it is not unthinkable that an intelligent and technological species should have a duration comparable with the timescale of cosmic phenomena and that many intelligent species may therefore exist at the same time. But the difficulty in imagining a very long-lasting intelligent species because of the continuous acceleration of technological progress might be simply due to the impossibility of imagining life in a context radically different from the one we know. Societies have always had the tendency to think they have reached the apex of human development, a typical example being the statements about the end of science that circulated among physicists in the last part of the nineteenth century. The idea that all there was to be discovered and invented was already at hand was so established that many scientists discouraged their best students from undertaking a scientific career. A Cro-magnon man was certainly unable to imagine life in an agricultural village and, just two centuries ago, the statement that only a small number of people would farm the land would have caused disbelief and raised many questions on what all the others would do.

EVOLUTION BEYOND HUMANS

The technological development of the various human species has always followed an S curve: a slow beginning, followed by a rapid increase, and then a stabilization (Figure 4.5). The sudden increase usually follows the passage from one species to the other or at least some strong somatic changes documented by fossils, particularly the increase in brain volume. Perhaps the only exception is the last of these expansive phases, the revolution of the Upper Paleolithic. Beginning about 100,000 years ago, a strong technological and cultural development occurred (but one ought perhaps to speak of the birth of *culture* as distinct from technology, more than of a cultural development) which was not accompanied by remarkable somatic changes. From *Homo sapiens praesapiens* and *Homo sapiens sapiens* (but also from pre-Neanderthal and Neanderthal men) onward, there are no physical changes remotely comparable with the cultural development.

This phase is still continuing today, with a constant acceleration, but since then modifications of the human species have been minimal. From this observation some deduce that, once the level of intelligence and consciousness of *Homo sapiens sapiens* is reached, physical evolution stops and further progress is only achieved through cultural evolution. A variation of this formulation is the hypothesis that further human evolution consists of the realization of a noosphere at the planetary level, which takes the form of a sort of collective intelligence, as if the highest possible encephalization in a single being has been reached in humans and further progress is possible only through a "network" (to use a computer analogy) of individual intelligences. This hypothesis is perhaps too teleological and thus unacceptable to those who, like the majority of biologists, think that evolution does not move in any particular direction, and that the tendency toward complexity is only the result of casual variations. Moreover, it is hard to see what kind of mechanism could allow the development of this collective intelligence.

Others think that this stability of the human species is just an error of perspective: *Homo sapiens sapiens* has existed for about 100,000 years, while the previous human species remained almost unchanged for much longer periods. Besides, they point out, no species has an indefinite or very long duration: either it evolves into some other species or it becomes extinct. The present phase of development will therefore slow down to a complete stabilization, until a new species will evolve from our species. It will produce a civilization that is totally unimaginable for us, just as our industrial civilization was certainly unimaginable for *Homo erectus*.

Finally, the idea that humans will give rise to more evolved organisms not through the usual evolutionary mechanisms but creating them directly with their intelligence must be mentioned. The basis for this idea is the opinion that humans will build more and more powerful computers, up to the so-called *Von Neumann machines*, that is, intelligent machines able to build their own copies. These universal self-replicating builders will become completely autonomous and will eventually replace humans and represent the next step in evolution. Until some years ago, it was actually thought that progress in the field of computers would replace an evolutionary line based on the biochemistry of carbon with one based on silicon chips. This trend could be implemented gradually, with the substitution of increasingly complex artificial organs into the human body. Artificial hearts, artificial eyes and ears, and mechanical limbs directly controlled by the brain are all being developed. In the future, these prostheses could be used not only to replace organs damaged by disease or accidents but also to improve human performance, possibly that of individuals who must operate in difficult environments. Science fiction described cyborgs, partly mechanical and partly biological beings, often in a disturbing or even apocalyptic way.

It is likely that in this field, too, the real world will go well beyond imagination—perhaps the very idea of cyborgs is already old. Recent developments in genetic engineering, nanotechnologies, and technologies based on a mixture of biology and traditional technology in general, together with a decrease of confidence in computer-based artificial intelligence, lead us to imagine different scenarios. The prostheses to substitute damaged organs or improve human abilities will not be mechanical but will more likely be biological, and the artificial beings that constitute the next step in our evolution will be living beings designed using the techniques of genetic engineering rather than machines controlled by intelligent computers.

Though none of these scenarios can be ruled out, it is more likely that if human evolution continues beyond the present stage, it will happen in the context of the human expansion in space.

Chapter 4

THE EXPANSION OF INTELLIGENT LIFE

Life spread on our planet to occupy every possible geographical site and ecological niche. Recent discoveries in meteorites of lunar or Martian origin and the possibility that very simple forms of life are transported from planet to planet in the Solar System, and perhaps even from one system to another, suggest that the tendency of life to spread is actually a more general phenomenon. But once an intelligent and technological species evolves on a planet, there is a faster and more efficient way to disseminate life beyond the planet and system of origin: the development of a technology allowing movement from planet to planet, transporting living beings and colonizing new habitats on an increasingly larger scale.

Recently a hypothetical principle has been proposed, namely the *conscious life expansion principle* (CLEP).[12] In its strong form it states: "An intelligent, self-conscious species evolving on a planet is eventually able to set about space exploration. This enterprise is neither an option nor a casual event in the species' history, but represents an obligatory way to diffuse high level life outside the place where it developed." For this principle to hold, the laws of physics must allow all forms of intelligent life to undertake interstellar travel; such a possibility must be given in any galaxy where intelligent life exists, independently from the number of inhabited systems at any one time. A weak form of the same principle exists, which states: "Universal laws are life oriented. As a very special case they allow conscious life to accomplish interstellar flight. Each civilization could be strongly motivated to either exploring the Universe without leaving its solar birth system or expanding to other star systems."

Expansion in space might then be a fundamental feature of intelligence and therefore also of hypothetical intelligent beings that may have evolved elsewhere, in our galaxy or in the Universe at large. In a sense, intelligence would be a prerequisite that life must develop for continuing its expansion out of its own planet of origin, just as it had to develop lungs (or any other organ able to obtain oxygen directly from the atmosphere) to colonize dry land, or wings to fly over the surface of the planet.

Perhaps this is too anthropomorphic a view, since mechanisms allowing nonintelligent or nonconscious beings to leave their planet of origin and

---

[12] G. Vulpetti, "On the Viability of Interstellar Flight," *49th Int. Astronautical Congress*, Melbourne, October 1998; G. Vulpetti, "Problems and Perspectives of Interstellar Exploration," *Journal of the British Interplanetary Society*, Special Issue: *Modern Views on Interstellar Flight and Related Key Disciplines*, September–October 1999. In this paper the problem of relationships between the CLEP and the anthropic principle in its various forms is also dealt with.

travel in space might be possible. Moreover, there is the possibility that living beings might exist whose original habitat is not a planet but space itself. As usual, we must refer to the only example of life that we know, life on Earth, and in this case it is clear that, apart from the possibility of microorganisms occasionally being transported by meteorites launched in space by cosmic collisions, intelligence and technology are essential for space flight. Some have objected to this principle because of the presumed difficulty or total impossibility of human expansion in space, particularly at interstellar distances. Many think that, at least in the case of the Solar System, the environments on all other celestial bodies are so hostile to life that space missions will be limited to sending humans into Earth orbit and to the automated exploration of the closest planets and their satellites.

While in the 1950s it was a commonplace that within a few decades humans would establish at least an outpost on the Moon and begin exploring Mars, only 20 years later these hopes were considered worthless dreams. This is partly due to the realization of the hostility of planetary environments. As mentioned in Chapter 3, this disappointment was mainly due to the results of research by the first probes that reached Mars and Venus. However, it is likely that deeper reasons are to be found in public opinion's change of attitude toward science and technology and in the disappointment that followed the disproportionate expectations placed on the *Apollo* missions. The results of the probes showed that some ideas of the habitability of Mars and Venus were groundless, but it was well known that the lunar environment was very hostile and the *Apollo* missions showed that humans are able to adapt themselves surprisingly well to work in these conditions. Lunar bases were therefore much closer to implementation in the 1970s than they were 20 years earlier, but plans to build them were only considered again in the 1990s.

In space, as well as on the Moon and on Mars, humans must live in artificial environments, but they won't appear subjectively more artificial than those we are accustomed to nowadays. A hotel room or an office with air conditioning are not much less artificial than a future lunar base; apart from the different technologies and the much higher costs, the subjective feelings of those living in them will not be much different. When traveling in an aircraft flying at the limits of the stratosphere, we are in an environment where unprotected human life is impossible: the difference between the environment surrounding an airliner and that surrounding a space station is not that meaningful from this point of view, as in both cases only an artificial environment allows the survival of complex life. Humans already live in an artificial environment, even if they are so accustomed to it that they often do not realize it is not natural. The environment in which terrestrial colonists will live on the Moon or on other celestial bodies won't be much more artificial, though perhaps they will need a long time to get used to it.

Many studies and detailed projects for large habitats in space and bases on the Moon and on Mars have been made: the technology is by now ready and there are no problems about their feasibility, even if many details still require studies and experiments. Although human expansion in the Solar System, or at least in the closer parts of it, does not require a technological revolution, only the political will to continue along the way already traced and studies aimed at reducing the costs, many think that automatic missions and, even more so, crewed ones beyond the limits of the Solar System imply difficulties and costs that make them almost impossible.

Such objections are based on the present state of science and technology and, if they may hold in the short period, they have little ground if they refer to a more distant future. Yet they can only be answered on the basis of historical precedents, that is, the many things that were thought to be impossible in the past and which later technological advancement made feasible, in fact within almost everybody's reach. The main objection to very long-range space travel is based on the enormous quantities of energy needed. To send an automatic probe to the planetary system of a nearby star with a speed allowing us to obtain scientific results within some tens of years, a quantity of energy must be spent comparable to the annual energetic budget of an industrialized country. Yet historical precedents show that the energy available to humanity grows constantly. A single airliner, for instance, uses in one year a quantity of energy larger than that used by a European country in the Middle Ages. Arguments of this kind do not really prove anything, but let us hope that what now seems a limitation we cannot overcome might in the future be mastered.

The most founded objection to the feasibility of interstellar travel concerns the long time needed for such journeys. Here the limitations are not due to present technology, but to a precise physical limit: any object or transmission containing information cannot move in space at a speed greater than the speed of light. The shortest travel time between two points in space, expressed in years, is therefore equal to the distance in light-years between the two points. And this is assuming that a technology allowing to overcome all the difficulties linked with the enormous quantities of energy needed for such a trip is available. Interstellar distances are so large that travel times of at least many years or even tens of years are required.

Trips of this duration rule out frequent voyages, particularly two-way travel, and complex relationships between persons and political entities located in different systems. Even simple exchanges of information can be only of historical and scientific type; that is, a community can be informed about the history of other communities living in nearby star systems and their scientific discoveries. Diplomatic and commercial relationships or the formation of political entities embracing several systems are probably impossible. The duration of interstellar journeys might be subjectively reduced by using techniques like hibernation, much employed by science fiction writers and today put on a scientific basis by recent studies on the hibernation of some animal species.[13] Moreover, if the speed is sufficiently close to the speed of light, time contracts and the time needed to reach distant star systems may be reduced to a few years or a few months. This consequence of the theory of relativity has been verified experimentally and, even if it appears to be a paradox, it is ascertained beyond any possible doubt. However, the quantity of energy needed for the journey increases greatly when approaching the speed of light and therefore the contraction of time requires a huge energy expenditure.

These considerations do not greatly modify the scenario seen earlier: whether travelers are hibernated or the relativistic contraction of time is exploited, travel will only be short for the people on board the starship, while for those who remain at home time will flow at the usual pace. In the case of two-way journeys, the travelers will find on their return that those who remained on Earth have grown old or, if the destination is at hundreds of light-years from Earth, have died centuries ago. Someone joked that the only reason for which someone might choose to embark on a two-way interstellar journey is to find his savings in the bank, when he got back, increased by orders of magnitude thanks to the interest accumulated in centuries, without having grown too old to spend it in the meantime!

But these considerations do not rule out the colonization of nearby star systems by an intelligent species, particularly if we look at things in a cosmological time-scale. A civilization able to travel at speeds close to the speed of light can cross our galaxy in about 100,000 years and reach many nearby galaxies within a few million years. A progressive colonization, starting from the closest systems and then including more and more distant stars is not therefore incompatible with science. It is possible to assert that a civilization really willing to colonize a galaxy, and able to travel at the maximum speed compatible with our scientific knowledge,

---

[13] T. Kondo, "Approaching Artificial Control of Hibernation," *3rd IAA Symposium on Realistic Near Term Advanced Space Missions*, Aosta, Italy, July 2000.

might succeed in its goal in about 10 million years. Such a time is enormous if compared to the scale of human life or of local history, but corresponds to the time taken by our species to evolve from the apes and is only one four-hundredth of the age of Earth. And the time needed for the colonization of the galaxy does not depend much on the duration of travel between one system and the next, but on the time needed for the colonies established on a planet to launch a new mission toward nearby stars; it is therefore not greatly influenced by the technology developed for interstellar travel or by the energetic and economic considerations above.

Interstellar colonization might be implemented using large spaceships that carry thousands of people, traveling at a relatively low speed (compared to the speed of light) for centuries. These *space arks* or *world ships* must be actual habitats, designed to host hundreds of thousand or even millions of people in space, in an environment similar to that on Earth.[14] The building of space arks does not go against any theoretical obstacle and does not require much progress in fundamental science, though the technological challenges and the investments needed are enormous. To reduce these problems, many alternatives have been imagined to avoid the need to build spaceships large enough to host a human community for many generations. Obviously it is possible to decrease the size of the starship by increasing the speed, thereby reducing travel time. If relativistic speeds are reached and the contraction of time is exploited, the size of the ship might be further decreased. If the duration of the trip, as felt by the travelers, is reduced to a few years or even to less than one year, travel will not be much longer, for instance, than the first human exploration journeys to Mars or many exploration journeys of the past.

As already mentioned, another solution is hibernation of the crew. We are close today to understanding the factors that allow many mammals to drastically slow down their metabolism to escape damage as they face the extremely unfavorable conditions encountered in the winter months in many regions of the Earth. This is not a hibernation obtained by cooling the body (something like deep-freezing the crew before departure and defrosting them upon arrival), but a biochemical process induced by hormones that cause a decrease of the pace of the metabolism and a lowering of the body temperature. There is no real reason why the same process allowing bears and squirrels to face hostile winter conditions might not allow humans to face the most hostile environment of all—space.

Another solution to the difficulties of interstellar colonization is to send

[14] G.K. O'Neil, *The High Frontier: Human Colonies in Space*, Mondadori, Milano, 1979.

automatic ships with frozen human embryos and machines capable of creating an environment favorable to human life on the destination planet. The same machines would then take care of the development of the embryos in artificial wombs and assist the newborn in their growth. This scenario presents serious obstacles, of a moral but also scientific and technological kind (think of the difficulties in ensuring normal psychological development for people raised in such conditions). The machines needed for an interstellar colonization of this type would have to be endowed with an extremely developed artificial intelligence, and there are doubts on the feasibility of such devices, even in the very long times involved in a hypothesis of this kind.

Those who believe that in the future it will be possible to build actually intelligent machines, think that expansion beyond the Solar System may occur using *Von Neumann probes*, space probes that, once landed on a celestial body like an asteroid or a rocky planet, are able to build copies of themselves. The strategy for interstellar exploration proposed for instance by physicist F.J. Tipler[15] is in principle quite simple: a Von Neumann probe is launched toward the nearest star using a suitable propulsion system, which may be of the present type or little more advanced. Actually, if it is not crucial to receive the results transmitted by the probe within the life span of the generation that launched it, long flight times may be acceptable. If some hundred thousand years are considered acceptable, the technology used for the *Voyager* probes would be suitable. Once the destination system is reached, the probe lands on a rocky celestial body and begins to replicate itself, building new probes that leave the system toward the nearest stars. After completing its primary task of continuing the expansion toward nearby systems, the probe starts performing its scientific work and sending the results back to Earth.

Tipler suggests that with a strategy of this kind and with a sufficiently advanced technology, exploration can proceed at a pace of about 10 light-years every 60 years. At that speed, the whole galaxy could be explored in about 600,000 years, an incredibly short time considering the enormous task. But his forecasts go well beyond that, as he states that a single intelligent species, by disseminating Von Neumann probes, is potentially capable of exploring the whole Universe. Such intelligent machines may do much more than perform exploration tasks, as it is possible to imagine them reproducing organic life. A species able to use Von Neumann probes could then colonize, occupy, and finally control the whole Universe!

This scenario aroused strong doubts, of a technological and more

---

[15] F.J. Tipler, *The Physics of Immortality*, Mondadori, Milano, 1994.

generally moral type. Even if we consider it morally acceptable to build intelligent machines, the idea of disseminating throughout the galaxy and then the Universe devices able to replicate themselves in an uncontrolled way raises strong opposition. Carl Sagan, for instance, firmly opposed this perspective, holding that a technological civilization must prevent the construction of interstellar Von Neumann machines and set strict limitations to their internal use.

Ethical doubts are strengthened by the fact that it is impossible to be sure that, after many replications, particularly where their programming is concerned, no errors would be introduced. It is possible to check the program and modify it by radio from Earth, but only for the first duplications, since distance in space and time quickly becomes so large that only onboard systems may later control the replication process. The same mechanism of mutations and natural selection that controls evolution on Earth would then start operating on Von Neumann machines. Machines programmed on Earth will slowly adapt to the places where they operate, and nobody can say what they may become once they stop behaving exactly as their builders had foreseen. Fortunately, the possibility of building an intelligent machine and, even more, a Von Neumann machine, is still remote; forecasts about artificial intelligence continue to slide further ahead in time and the increase in computer performances have not really brought us any closer to that goal. Many scientists doubt that building such a machine is altogether possible, at least using present technologies or those predictable for the future.

If all the above scenarios deal with a distant future, it is already possible to spread life in the Universe today by sending microorganisms to other star systems. As already mentioned when speaking of the possible diffusion of life from planet to planet, some microorganisms can survive for tens or hundreds of thousand years in space and even in the interstellar environment. The use of propulsion devices capable of reaching interstellar destinations within a short time is then not needed and even a probe like the *Voyager* could be used, since it might transport many kilograms of microorganisms and disseminate them in space close to the destination star. They would then be gravitationally captured by the planets of the system and, after falling on them, start the evolution of life-forms on every planet where the environment allows it.

A scientific association has been created, the Society for the Interstellar Propagation of Life, with the aim of promoting the dissemination of life in nearby systems, the proposed means usually being very thin, light, and relatively cheap solar sails. A swarm of such sails could bring an enormous number of microorganisms, able to colonize all the planets of a star system with characteristics suitable for life to take root. This artificial panspermia is nothing more than the space version of the method used on Earth by many plants that release their seeds in the wind or by coconut palms, whose fruits brought by the sea disseminated that species throughout the islands of the Pacific Ocean.

Yet this way of diffusing life cannot be controlled and the microorganisms launched toward other systems, besides spreading life based on DNA, could cause the extinction of other forms of life, different and perhaps even more evolved. Inseminating only very young star systems or even stars still in their formation stages, where life has not yet started to develop, would avoid causing the extinction of other life-forms but could prevent them from developing. Humankind has many alternatives to spreading life on other planets, first in the Solar System and then in other systems where life has not yet started. These are, however, general aims and hold for any intelligent species that evolved in our galaxy, or elsewhere, so that any technologically advanced species might create a sort of zone of influence around its own system, colonized by life-forms deriving from those of the planet of origin. Besides, it is reasonable to expect that the extension of such zones of influence is larger for the older intelligent species.

It is natural that a scenario of this type would breed situations of conflict when two spheres of influence come into contact with each other or when one of the spheres reaches a planet on which some forms of life developed autonomously. If the expansion is managed directly by the intelligent species that promoted it, ethical problems similar to those encountered on Earth when different cultures come into contact, will arise, be amplified, and made more difficult by incomparably larger differences. Situations could be imagined in which it is difficult even for one or both the intelligences to recognize the other life-form as also intelligent. There might be cases in which the definitions of life or intelligence are so different as not to include the counterpart.

If the expansion happens without the direct participation of the intelligent species, as in the case of panspermia, the dangers may be even greater, though fears of biological contamination are probably much exaggerated. If an alien life-form arrives on an already inhabited planet it will be at a disadvantage in comparison to the native forms of life that had already adapted to the environment, and the new form could hardly represent an actual danger for them.

It is possible that the expansion of life on other celestial bodies finds strong limitations in the difficulty, or impossibility, of finding suitable environments. The forms of higher life that developed on Earth after

163

unicellular organisms had deeply transformed its atmosphere cannot live on a planet that never hosted life, if not in a protected environments. In a similar way, it is possible that the process of mutual adaptation of a planet and the living beings on it is such that it is impossible for a species to survive, without suitable protection, on a planet different from that of its origin. This does not prevent the colonization of other celestial bodies, but compels the colonists to create protected environments, separated from that prevailing on the planet.

A sufficiently advanced technology allows the environment to be transformed to make it suitable for hosting life from the colonizing planet: in the case of humans from planet Earth this process is referred to as "terraforming," a term introduced by science fiction but now accepted in scientific language. If the planet does not host any life, this process will not present significant ethical problems, while in the opposite case it could easily cause the extinction of the native forms of life.

THE SEARCH FOR INTELLIGENT SIGNALS

The only known example of the evolution of a form of intelligent life— our own—has so far been considered, with attempts to predict its future developments, with the main purpose of drawing information that could be useful in assessing the possibility of the existence of extraterrestrial intelligence. It is extremely unlikely that extraterrestrial intelligent beings exist in the Solar System. It would be an exciting result just to discover a living being similar to our bacteria. While nothing can be excluded with certainty, since in theory intelligent life-forms might exist even on Jupiter or even on Venus (clearly they would be extremely different from the life we know), if there is to be some probability of success we must look for intelligence beyond our Solar System.

Since we are not able at present to study extrasolar planets in detail, to send probes or to travel through interstellar space, the only possible way to discover other intelligent species is to receive messages they may broadcast into interstellar space, either on purpose or as an unwanted result of their activity, or in some way to observe signs of their presence. It is not unreasonable to think that it is possible to discover other intelligent species that betray their presence with activities detectable from great distances or that intentionally try to communicate with other species. At our technological level, a search of this type can be performed by studying the electromagnetic waves reaching us from space, mainly in the visible or infrared region of the spectrum or in that of radio waves. So far, the search for intelligent signals mainly concentrated on the attempt to receive radio signals, in the belief that this type of transmission is the simplest and most effective means of communication at interstellar distances.

The scientific search for forms of intelligent life is known as SETI (Search for ExtraTerrestrial Intelligence), and although, as mentioned in Chapter 1, more ancient precedents exist, its beginning is usually traced back to a famous paper published by Philip Morrison and Joseph Cocconi, entitled "Search for Interstellar Communications," in the journal *Nature* in 1959. The authors asserted that the radiotelescopes existing on Earth were able to detect possible transmissions broadcast, purposely or not, by hypothetical civilizations located on planets orbiting many stars, even tens of light-years distant from us. Accordingly, they concluded that it was possible to bring onto solid experimental ground a problem that had earlier been dealt with more as philosophical than scientific. Radioastronomers suddenly found themselves on the front line in a research field that, up to that point, was completely outside their scientific interests.

One attempt to put the search for extraterrestrial intelligence on a rational basis by estimating the number (N) of technological civilizations that perform activities we can identify existing in our galaxy at present[16] is the equation introduced by Frank Drake. It can be written in the form[17]

$$N = R \times f_p \times n_e \times f_l \times f_i \times f_c \times L$$

where R is the number of stars suitable to sustaining life (generally interpreted as stars of solar type) that form every year in our galaxy, L is the time for which a civilization is able to communicate with us, n_e is the number of planets at a distance from their sun allowing the formation of life or, in general, the number of planets suitable to life in a system, and f_p, f_l, f_i, and f_c are coefficients smaller than unity expressing the astronomic, biological, and cultural factors that influence the phenomenon:

- f_p is the fraction of stars of solar type that have planets;
- f_l is the fraction of planets on which life could develop, that indeed host living beings;

---

[16] The term "at present" must be referred to as the time needed for a signal coming from a distant civilization to reach us. A civilization "at present" existing at a distance of 2000 light-years, actually existed 2000 years ago.

[17] The Drake equation has been written in various different forms, all essentially equivalent.

- f_i is the fraction of planets on which life exists, that indeed host intelligent beings;
- f_c is the fraction of civilizations that develop a technology detectable by us.

The Drake equation allows us to rationalize the problem, but does not supply a numerical result, since some of the coefficients can only be evaluated with large margins of uncertainty. According to the assumptions we make, the Drake equation could give results that span from 1 (we are the only beings able to communicate in the galaxy) to hundreds of millions (the galaxy swarms with beings trying to communicate with each other).

Certainly, f_i could be close to 1, since life on Earth developed very quickly, as if it were almost an automatic process. f_p and n_e could be higher than was thought until recently, since many extrasolar planets (not of terrestrial type, however) were found and we now realize that liquid water may exist at distances from the Sun very different from those previously assumed.

R also depends on whether double stars may have a planetary system: in this case, its value would be higher than expected. To only consider stars of the solar type may be too conservative, since nobody is sure that life cannot develop at a suitable distance from a red giant or from a star of other types.

The value of coefficient f_c is very uncertain, since it depends on objective considerations such as the technical ability to communicate, but also from choices that, dealing with intelligent life, must be considered as subject to free will: a civilization, able to communicate, might not do it for indifference, fear, xenophobia or any other reason.

The most uncertain term is perhaps L, not only because of doubts about the duration of an intelligent species at the stage in which its technology allows it to send messages. The simplest reason for which an intelligent species may stop transmitting radio signals is not that it has destroyed itself, but that it has moved beyond the technological stage in which it uses radio waves to send information. On Earth we are already replacing television broadcasting with cable TV, which obviously cannot be received by outsiders. Today the most powerful transmitting stations are military and scientific radars, but it is possible that within a relatively short time (short when measured on the time scale of cosmic phenomena) they will also be replaced by devices working on different principles. There are already some signs in this direction: very long-range communications with space probes may in the future use laser light instead of radio waves. The fact that a laser beam is much better focused than a radio beam, makes a laser a much more efficient means of transporting information, but it also makes it much more difficult for those who are searching for signs of intelligent life to detect it.

If the stage at which an intelligent species uses radio waves to exchange information lasts only some hundred years, the possibility that several species sending radio messages exist at the same time at a distance allowing them to communicate is almost nil.

To get an idea of the results that may be obtained from Drake's equation, assume that:

- $R = 10$ stars per year (every year 10 stars suitable for life will form in our galaxy);
- $f_p = 0.1$ (10 percent of stars have some planets)—an arbitrary but reasonable value, owing to the number of planetary systems discovered;
- $n_e = 3$ (in every planetary system there are three planets in a zone suitable for life)—a value based on the Solar System, in which there are three celestial bodies that could potentially host life: the Earth, Europa, and Titan. Mars could also be added to the list, but the two satellites of Jupiter and Saturn could be eliminated;
- $f_l = 1$ (all the planets in a zone suitable for life end up hosting life)— certainly an upper limit that might be extremely optimistic. Together with the estimate of n_e seen above, it implies that Europa and Titan are (or better, have been in the past or will be in the future) inhabited;
- $f_i = 1$ (all the planets on which life begins will end up developing intelligent beings)—also an upper limit, likely incompatible with the estimates of n_e and of f_l. In particular, Europa and Titan hosted or will host intelligent beings;
- $f_c = 1$ (all intelligent beings eventually develop a technological civilization)—an upper limit;
- $L = 10,000$ (the time for which a civilization is able to communicate, or at least is made visible by its activities, is equal to 10,000 years).

With these values for the parameters, Drake's equation yields a number of civilizations in our galaxy able to contact us, $N=30,000$. With this kind of result we should expect the average distance between civilizations that broadcast to be about 700 light-years. Note that this estimate is really an upper limit, since many coefficients have been largely overestimated. Only for L could a much larger value (even by some orders of magnitude) have been used, since a technological civilization might last much longer, but on the other hand it is doubtful that a civilization would use a technology detectable by us for such a long time.

Now assume that, without changing the values of R (10) and of f_p (0.1), we set

- $n_e = 1$ (in every planetary system there is just one planet in a zone suitable for life)—a value that for the Solar System derives from the traditional definition of habitable zone. Perhaps it is even too large, in light of the recent discoveries about the orbits of extrasolar planets;
- $f_l = 1$ (all the planets in a zone suitable for life end up hosting it)—a value that may be justified by the quickness with which life developed on Earth;
- $f_i = 0.01$ (only 1 percent of the planets on which life begins end up developing intelligent beings)—a low value, but based on the time needed by evolution on Earth and on the many catastrophes that could halt it;
- $f_c = 0.5$ (half of the civilizations develop technology);
- $L = 200$ (a civilization only spends 200 years using radio waves to communicate).

These values yield $N = 1$, that is, there is only one technological civilization in our galaxy—us.

If N is low, we can only give a statistical interpretation: it has no meaning to expect, for instance, that if $N=1$ there is always someone broadcasting messages, since there will be periods in which there are many transmissions and others in which nobody sends signals. Besides, we could obtain much lower estimates, that would give an infinitesimal probability of receiving messages.

Since it is unlikely that civilizations much more advanced than ours use radio waves, it might be useful, as well as searching for radio transmissions, to search for traces of very large-scale engineering works performed by extremely advanced civilizations. Physicist Freeman Dyson, for instance, suggested that a civilization exploiting all the energy produced by the star around which its planet orbits may build a thin spherical structure, using material obtained from asteroids and others small celestial bodies. Such gigantic objects, usually referred to as Dyson spheres, might have a radius of some astronomical units, and include the star and all inner planets, intercepting all the energy produced inside its surface.

The laws of thermodynamics state that a Dyson sphere should send out part of the energy it collects in the form of heat at a fairly low temperature, therefore as infrared radiation, detectable even at a very large distance. Detailed observations have been performed analysing the radiation of 54 solar-type stars to look for anomalous emissions in the infrared region of the spectrum, without obtaining any evidence of Dyson spheres. Certainly 54 stars out of billions of possible candidates are very few and the large astroengineering works that a very advanced civilization might undertake are not limited to Dyson spheres. In this case the acronym SETA (Search for ExtraTerrestrial Artifacts) might be more appropriate than SETI.

The Russian radioastronomer Nikolai Kardashev suggested that very advanced technological civilizations be recorded in three classes: in the first are those that use an average power equal to all the power that their planet receives from the star around which it orbits; in the second those that use a power equal to all the power produced by their star; and finally, in the third, those that use a power equal to that of all the stars in their galaxy. Civilizations of the second and the third types, provided they exist, should be easily detectable from the traces of their activity.

Studies focused on the analysis of the visible and infrared regions of the electromagnetic spectrum are under way, yet there is no doubt that the majority of the SETI studies are performed in the region of radio waves. A survey of the research projects presently active is reported in Appendix B.

The research may be based on two distinct approaches: a study of a certain number of promising objects or an investigation with a global coverage of the sky.[18] In the first case the radiotelescope is aimed at a particular star, chosen on the basis of a series of considerations, such as similarity with our Sun, age, distance, and the existence of planets. Alpha Centauri, for instance, might be an interesting target, owing to its closeness, that makes easy to receive unintentional transmissions leaking from a possible civilization and the circumstance that one of the components of this multiple star is almost identical to the Sun. Unfortunately, however, it is a triple star and therefore the existence of planets in orbits allowing the development of life is rather problematic. The Barnard Star, at a distance of 6 light-years, was thought for a long time to have a planetary system, but today few astronomers would endorse this statement. At any rate, it is thought to be too old to host an intelligent species.

The most interesting stars in our galactic neighborhood are Epsilon Eridani, a young star slightly smaller than the Sun, at 10.7 light-years, and above all Tau Ceti, a star similar to the Sun, of the same age and located at only 12 light-years. These two stars have been observed for a long time and the first one caused the first false alarm when, during one of the first attempts of the project Ozma, Frank Drake received a pulsed signal, very clear and distinct, immediately after aiming the radiotelescope in its direction. But the signal was not repeated, and after quite a long time the

---

[18] For a detailed description of SETI activities and equipment see, for instance, J. Heidmann, *Extraterrestrials*, Cambridge University Press, Cambridge, 1996; and G.A. Lemarchand, *El llamado de las estrellas*, Lugar Cientifico, Buenos Aires, 1991.

scientist discovered that it was probably caused by an aircraft flying at high altitude. As already mentioned, Epsilon Eridani has a giant planet orbiting it at about 3 AU. Unfortunately, the orbit of the planet is very elliptical, which makes the existence of habitable planets very problematic.

When performing a total coverage search, the whole celestial sphere is slowly scanned by the radiotelescope in the hope of finding a signal. It is not limited to observations in the directions in which stars are located, since it cannot be excluded that a source of transmissions might be found in space, far from any celestial body. We might, for example, intercept the transmission of a spaceship or, more simply, of an alien automatic probe, actually aimed at the planet of origin or at another space vehicle.

A third strategy is becoming more and more popular and is substituting for the two mentioned above: the study of the recordings of radio-astronomic signals performed for other purposes, in the hope of finding signals of artificial origin. Clearly the probability of finding an intelligent signal in this way is lower than that of finding it through a purposely designed search, since the most interesting targets for other radio-astronomic studies are different from those of interest for SETI and also because optimal observation procedures are different, but the cost of this way of performing SETI studies is so low that the total number of observations may be increased by orders of magnitude. When one has no clear idea of where to look and what to look for, it may be better to deal with many observations performed in nonoptimal conditions than to use a small number of very good observations.

Actually, neither the origin nor the frequency of a possible signal is known. Even if we only consider the frequency range of radio waves, the possible channels are many, particularly as the main characteristic of an artificial signal is to be a narrow band signal, that is, a signal in which most of the energy is concentrated in a very narrow frequency range. Since the interstellar medium causes a dispersion in the frequency of radio waves, there is a limit to how narrow the bandwidth of the signal can be; nevertheless, the radio wave region of the spectrum contains billions of possible channels.

Many hypotheses have been advanced on the frequencies that could be preferentially chosen for interstellar communications. There is a rather large region in which the background noise is relatively low, spanning from slightly less than 1 GHz to about 100 GHz (Figure 4.6). Below 1 GHz the noise produced by our galaxy (mainly due to synchrotron radiation) becomes the limiting factor. This noise is lowest in a direction perpendicular to the plane of the galaxy ($b = 90°$ in the figure) and highest at low galactic latitudes, particularly in the direction of the center. At high frequency the limiting factor is quantum noise. In the above-mentioned frequency range, noise is mostly due to the background radiation at 3 K. This, however, only holds for an antenna located in space; when the receiver is on the surface of Earth, it is necessary to account for atmospheric absorption, particularly at the frequencies of water vapor (22 GHz) and oxygen (60 GHz). This effect, limiting the usable frequency range for an antenna located on Earth to between 1 and 10 GHz, is reported in the form of additional noise in Figure 4.6.

If someone is broadcasting with the aim of being received, that person will most likely use a frequency located in a range in which noise is low and, if trying to communicate with someone on a planet with an atmosphere containing water and oxygen (that is, trying to communicate with a terrestrial type life-form), will also take this effect into account.

Starting from the first paper by Cocconi and Morrison, a particularly favorable frequency has been identified, namely that of atomic hydrogen (1.442 GHz), since it corresponds to a zone where noise is very low and is a symbolic frequency that any being with a minimal knowledge of physics cannot ignore. Actually many other frequencies might be chosen, like that of the OH radical, and it is impossible to choose a single particular frequency in the rather ample field spanning a few gigahertz. The frequency range spanning the frequencies of hydrogen and that of the OH

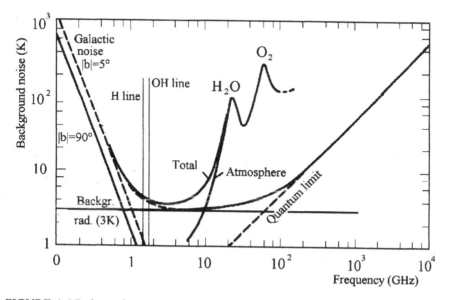

FIGURE 4.6 Background noise at various frequencies. The effect of absorption due to the atmosphere is shown in terms of noise. The values reported are average values, since the atmospheric effect depends on many factors, one of which is the direction in which the antenna is aimed.

radical is often referred to as the *water hole*, making a kind of analogy between the frequency range at which the various civilizations of the galaxy could choose to communicate and the ponds full of water around which animals meet in the grassland.

Since we have no idea of the frequency at which the hypothetical extraterrestrials may broadcast, we must use receivers that are able to receive many channels simultaneously, or elaborate wide bandwidth signals received by radiotelescopes in such a way as to separate the various channels with the aim of studying them one by one. In the first case, special receivers designed for this type of research are available today, and can deal with tens of millions of channels. In the second case, computers with enormous power are needed, especially if completely automatic signal processing is required. Actually the calculations to be performed are not very complex, and instead of a very powerful computer it is possible to use many computers with a relatively low performance. This is the way followed by the SETI@home project—a part of the SERENDIP project implemented by the University of California at Berkeley using the large Puerto Rico radiotelescope. It supplies software and sets of data to be analyzed through the Internet to many hundreds of thousands of volunteers worldwide. The analysis of the signals is performed by the computers in the time left over from their other operations; instead of visualizing a screen saver, millions of computers worldwide, particularly in universities and private homes, every day scan signals from space, searching for traces of intelligence.[19] At present (December 2005), having reached more than 5.5 million connected computers in 226 countries, SETI@home is going off line and migrating to a new infrastructure, the Berkeley Open Infrastructure for Network Computing (BOINC). The new platform will allow SETI@home to use data beamed directly from different telescopes, including those in the Southern Hemisphere, and to look at a wider radiofrequency range.

Besides the difficulties deriving from a complete ignorance of the frequency at which possible transmissions might be broadcast, there are the problems of the duration in time of the signal and of the possibility of obtaining confirmation of the discovery. Since the Earth and the planet from which the transmission comes rotate and move in a complicated way, the length of time for which the receiving antenna is in the beam may be very short and the signal might never be received again. This is naturally true for very directional signals unintentionally transmitted into

---

[19] The programs and the data can be unloaded from the site http://setiathome.ssl.berkeley. edu/.

space, whereas nondirectional transmissions such as our television broadcasts could be received for much longer periods, although necessarily being much weaker. Signals purposely broadcast with the aim of communicating with us should also be received for longer times, since whoever sends them should be aware of these problems and at least compensate for the motion of the planet of origin by choosing a not too directional transmission.

At any rate, the possibility of obtaining a confirmation of the discovery is fundamental, and little credit can be given to a contact of very short duration that is not repeated; there is a great danger of interpreting interference from a source located on Earth or in its proximity as an extraterrestrial signal, or even of being fooled by a hoax. (It would not be the first time that fraud or a joke has been at the root of a "sensational" scientific discovery.)

No doubt this research is extremely demanding and might last decades, despite the progress in the receivers and in computers to analyze the signals. Many radioastronomic observatories and public and private research organizations in many countries are deeply involved in it; it is often said that it is like searching for a needle in a haystack of cosmic dimensions. Wide international participation is needed to keep the whole sky constantly under observation at the various latitudes and to guarantee continuous coverage of any possible source during the whole day. Funding comes from public organizations, foundations, and private associations. For instance, in the United States, SETI received significant funding from NASA, and a very intense program had been launched. However, in 1981, the Senate, under pressure from Senator William Proxmire of Wisconsin, completely canceled the funds and went on to forbid NASA any activity in this field; the search was thus continued entirely with private funds, thanks to organizations like the SETI Institute and the SETI League. Currently the veto is somewhat less strict, but research is still mostly conducted with private funding. In the other nations the funds come from sources similar to those typical of astronomical and radioastronomical research.

Many wonder whether it is wise to spend large sums on research that has a low probability of success, at least in the short term. In this case the low probability of success is accompanied by the extreme importance of a positive result, and there is wide agreement in the scientific community to go on—more so as the amount of money involved is not very large, like that required, for instance, for fundamental physics research or for space exploration. A sense of urgency can often be felt in the words of those working in the field of SETI: the expanding telecommunication business, in particular the increasing number of telecommunication satellites, allows us to predict that in the future it will be more and more difficult to perform radioastronomic studies from Earth's surface, particularly when extremely weak signals must be studied, as in the case of SETI. This was the opinion of Jean Heidmann, for instance, who devoted the last years of his life to the project of a radioastronomic observatory located on the far side of the Moon, the only place completely free from Earth's electromagnetic pollution.

Observation from space, or at least using an antenna located at a sufficiently high altitude, is useful, since it avoids the signals being absorbed by Earth's atmosphere, allowing the study of the frequency range between 10 and 100 GHz (Figure 4.6). Radioastronomic satellites can solve this problem, but not those related to the noise due to human activity. An astronomic and radioastronomic station on the far side of the Moon represents an ideal solution, provided that humans do not in the meantime pollute even that uncontaminated environment with radio-waves. Actually, a suitable choice of site can limit the zone to be protected to a relatively small area, in order to avoid interference with other scientific and economic utilizations of the lunar surface: it is better to protect a small zone on the whole frequency range, rather than to extend the ban to the whole far side of the Moon, limiting it to a few selected frequencies.

For this reason, some radioastronomers proposed a treaty among powers with a sufficiently advanced space technology to ban all forms of radio transmissions from artificial satellites of the Moon and lunar bases that could interest the selected zone. The most suitable place seems to be the Saha crater, located slightly south of our satellite's equator (at exactly 2°S) and at a longitude of about 102°E. It is a circular crater with a diameter of about 100 km, surrounded by mountains 3000 m high, protecting the inner surface from short-wave and microwave transmissions from the surface of the Moon or from space vehicles not too high in the sky. It is never in sight of Earth, even in the case of 7-degree librations, but its closeness to the near side allows the installation of a permanent connection, through a 350-km cable or a few laser relay stations, with a radio station in the eastern zone of the Mare Smithii[20] (Figure 4.7).

At the start it will be just a small, automatic station. A single vehicle

[20] After the death of Jean Heidmann the project was reviewed and a new location suggested. The author, however, still supports the project in its initial form and thinks the location at Saha crater is a very good choice.

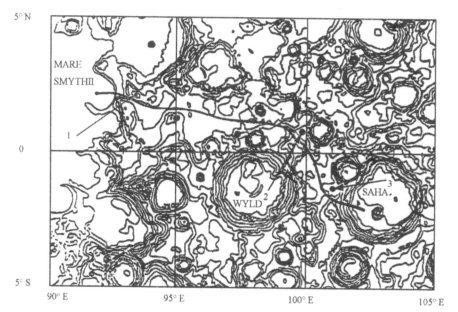

FIGURE 4.7 *Map of the lunar territory including the Saha crater and the eastern end of the Mare Smithii. The continuous line is the shortest route connecting the two places; the points marked 1, 2, and 3 are the sites in which three laser relay stations may be located to implement a laser link between the radiotelescope and the radio station that maintains contact with Earth.*

containing five landers could be launched from Earth. The largest lander, carrying the actual radiotelescope, will land in the Saha crater. Three other small landers, each one carrying a laser relay station, landing at locations marked 1, 2, and 3 on the map in Figure 4.7, and later being accurately positioned, constitute a laser telecommunication link. Finally the fifth one, with a radio station, will land in the Mare Smithii to allow communications with Earth. The radiotelescope will send the signals it detects and receive control commands without producing disturbances that could hamper its work. The radioastronomic station on the far side of the Moon may subsequently be enlarged. With a permanent lunar base, it will be possible to transform what was initially a small automatic station into a real observatory, if not permanently manned, at least visited from time to time by astronauts and scientists who will perform the necessary maintenance and upgrading operations.

THE RIO SCALE

At the International Astronautical Congress of 2000 in Rio de Janeiro, Ivan Almár and Jill Tarter proposed a scale[21] to evaluate the impact of a SETI discovery on society, not unlike the Torino scale for the danger of asteroid collisions. The scale is a dynamic one (in this, too, it is similar to the Torino scale) in the sense that a value is assigned when a detection is first announced, and it is then changed as the news is checked and new evidence accumulates, until a final value is obtained. The scale is based on four indices. The first three parameters are

- Q_1: class of phenomenon
- Q_2: type of discovery, and
- Q_3: distance.

The values of these parameters are shown in Table 4.1. They add up to give the parameter

$$Q = Q_1 + Q_2 + Q_3$$

which can take a value from 3 to 15. The fourth index is the credibility index (Table 4.2) δ, which has a value between 0 and 4/6.

TABLE 4.1 Values of coefficients Q_1, Q_2 and Q_3

Class of phenomenon	Q_1	Type of discovery	Q_2	Distance	Q_3
Earth-specific message	6				
Omnidirectional message	5	Result of SETI/SETA —steady	5		
Earth-specific beacon	4	Result of other kind of observation—steady	4	Within our Solar System	4
Omnidirectional beacon	3	Result of SETI /SETA —transient	3	Within 50 light-years	3
Leakage radiation	2	Result of other kind of observation—transient	2	Within our galaxy	2
Traces of astro-engineering	1	Reevaluation of archival data	1	Extragalactic	1

[21] I. Almár and J. Tarter, "The Discovery of ETI as a High-Consequence, Low-Probability Event," *51st International Astronautical Congress*, Rio de Janeiro, October 2000; I. Almár, "How the Rio Scale should Be Improved," *52nd International Astronautical Congress*, Toulouse, September 2001.

TABLE 4.2 *Values of the credibility index* δ

	δ
Absolutely reliable, without any doubt	4/6
Very probable, with verification already carried out	3/6
Possible, but should be verified before taken seriously	2/6
Very uncertain, but worthy of verification efforts	1/6
Obviously fake or fraudulent	0

The rating RS of the importance of a detection is the product of index Q times the credibility index δ

$$RS = Q \times \delta$$

The rating is then a number from 0 (event of no importance) to 10 (extraordinary event), as shown in Table 4.3.

The proponents tried to apply the Rio scale to various scenarios, as depicted by science-fiction movies and based on some false alarms and hoaxes of the past, with reasonable results: in science fiction, events usually have little uncertainty, and the rating quickly went up to 8 or 10 depending on the class of phenomenon (Q_1), while in false alarms and hoaxes the index quickly converged to 0, as it should.

TABLE 4.3 *Level of importance RS of a SETI event in the Rio Scale*

RS	Importance	RS	Importance
0	None	6	Noteworthy
1	Insignificant	7	High
2	Low	8	Far-reaching
3	Minor	9	Outstanding
4	Moderate	10	Extraordinary
5	Intermediate		

THE PROBLEM OF THE ANSWER

Although no messages of an artificial origin have so far been received, a few false alarms did occur in the past. Actually it is not at all easy to recognize a signal betraying the existence of an extraterrestrial civilization amid the clutter of transmissions, both natural and artificial, received by radiotelescopes. The first possible source of false alarms is transmissions due to human activities; they have all the characteristics that identify them as artificial and if their terrestrial origin is not immediately clear, misunderstandings of all sorts are possible. When the signals come from a transmitter located on Earth's surface, flying in the atmosphere or in low orbit, it is sufficient to compare the signals received by antennas located far from each other. If the signals come from distant space vehicles, things are more difficult and a deeper study may be needed. Since there are not that many vehicles in the Solar System, it is possible to compare the signals and the position of the transmitting antenna with the data on the few probes that are active to avoid all possible false alarms.

The natural signals commonly studied by radioastronomy can hardly be mistaken for artificial signals, but problems may be caused by signals that are identified for the first time. The most famous case was that of pulsars. When they were first discovered the certainty spread, for a time, that an artificial transmission had been received and to this day these sources are catalogued as LGM followed by a number: LGM ironically stands for *little green men.*

To avoid false alarms, the discoverers of a signal of potential interest for SETI must follow agreed procedures. This agreement, usually referred to as *post-detections protocols*, is reported in Appendix C. It is important to avoid false alarms as they would undermine the credibility of the research, besides upsetting the public. Whoever detects a signal must first avoid premature leakage of information; instead, he or she must inform other researchers, who will carry out accurate verifications. This procedure is essential: only the detection of the same signal by radiotelescopes distant from each other allows signals of terrestrial origin to be distinguished from those coming from space. But in this way it becomes impossible to take into account signals received only for a short period of time, as for instance highly directional beacons that accidentally cross the orbit of Earth, or signals that broadcast only for short intervals. A radar beam from Earth would be seen in this way from an extrasolar planet, and extraterrestrial scientists who adopt the same principle would never detect it or understand that it is a sign of our presence. Unfortunately, nonrepeatable events cannot be considered by science, whose method requires an experiment to be repeated and verified by independent observers.

The discovery of extraterrestrial intelligence is an event of exceptional relevance and the principle synthesized in the sentence "exceptional discoveries require exceptional evidence" must be applied to it. When performing research at the frontiers of scientific knowledge that involve themes of great impact on our view of the world, it is essential that the evidence supplied by discoverers is absolutely certain and unassailable.

Those who criticize this formulation say that this excessively conservative attitude results in a slowing down of scientific progress and might prevent some discoveries from being made. It is a motivated concern and the risk is real, but the opposite danger, that of accepting as evidence simple clues and building scientific theories on insecure bases, is far worse. Moreover, in fields like this, in which there is strong emotional involvement of the researchers, there is a risk of confusing one's desires or fears with scientifically ascertained facts and losing contact with the real world. The criterion of exceptional evidence—in the sense of unassailable verification confirmed by cross-checking performed by other researchers—cannot therefore be abandoned, even if, as some rightly fear, it might delay a discovery.

Only after it has been ascertained that the message is really evidence of the existence of extraterrestrial intelligence must public opinion be informed, in the forms decided by the top echelons of the United Nations. Once proof has been obtained that there is someone else in the Universe and that they are possibly looking for contact, the problem will be faced of whether an answer should be sent. Of course, the problem of whether transmissions should be broadcast is already present before having obtained a contact. Currently the tendency is not to broadcast any message and consequently what had been defined as CETI (Communication with Extraterrestrial Intelligence) is today known with the acronym SETI, where the word *communication* has been replaced by *search*, which does not imply transmission of messages on our part. Transmissions have been sent in the past, like that broadcast by Frank Drake from Arecibo on the occasion of the reopening of the radiotelescope in 1974, toward nebula M13 in the constellation of Hercules (Figure 4.8), but they aroused strong criticism and have been suspended.

Even stronger was the criticism against the transmissions broadcast by a Houston-based firm, Encounter 2001, using the Evpatoyia radiotelescope in the Ukraine. A first message was sent out in May 1999, with some scientific content together with the names and short personal messages from the company's customers. Many find it unacceptable that private operators send messages, for scientific or commercial reasons, which, if received by an alien civilization, may be interpreted as an official calling card from humanity.

But there is more. Those who think that it is unwise to broadcast transmissions into space say that it is potentially dangerous to let one's presence and position be known to beings we know nothing about. As we will see later, it is quite likely that the extraterrestrials with whom we could come in contact are older than we are, at least as intelligent a

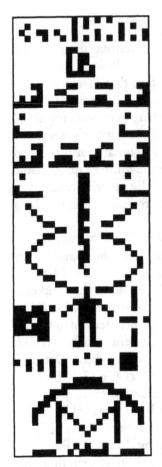

FIGURE 4.8 Graphic representation of the message transmitted toward nebula M13 in the Hercules constellation by Frank Drake using the Arecibo radiotelescope on the occasion of its reopening in 1974.

species, and much more advanced technologically: historical experience teaches us that every time a civilization came in contact with another that is technologically more advanced, the consequences for the first were very serious, in some cases to the point that it physically disappeared or was the victim of genocide. This consideration alone should suggest the greatest caution.

Some, on the contrary, think that contact with a civilization much more advanced than ours is free from danger, since such an advanced civilization must necessarily be more ethically advanced, too. In this case we would have nothing to fear from them, while contact would allow us to learn much, and not only in the scientific field. This formulation is made explicit with statements of various kinds, among which is the hypothesis that if a civilization is very aggressive, it will sooner or later destroy itself. As a consequence, a species cannot be very old and very aggressive at the same time: if they are more advanced than we are, they necessarily must be pacific. To this a further consideration is added: interstellar distances are so large as to make any danger very limited.

These considerations are not just influenced by an all-too-anthropomorphic approach; they are also influenced by the present situation and recent past of our own civilization, particularly by the Cold War and the fear that nuclear war might cause the extinction of our species.

THE SAN MARINO SCALE

As stated above, it is potentially dangerous to "give away our position" through transmission and it is a good idea to try to quantify such a danger so as to regulate the whole matter. With this aim in mind, Ivan Almár

TABLE 4.4 *Intensity index: I_0 is the current average background intensity of terrestrial noise in the frequency band of the transmission*

Intensity of the transmission	I
$\geq 100,000\ I_0$	5
$10,000\ I_0$	4
$1,000\ I_0$	3
$100\ I_0$	2
$10\ I_0$	1

TABLE 4.5 *Character index*

Character of the transmission	C
Reply to an extraterrestrial signal or message (if they are not aware of us yet)	5
Continuous omnidirectional, broadband transmission of a message to ETI	4
Special signal in a preselected direction at a preselected time in order to draw the attention of ET astronomers	3
Message with intention to reach ETI—at arbitrary directions for minutes or hours	2
A beacon without any message, e.g., a planetary radar	1

proposed another scale,[22] working in a similar way to the Rio scale to evaluate the danger linked with an interstellar transmissions from our planet. The *San Marino Index* (SMI) is the sum of two indices, I (intensity, Table 4.4) and C (character, Table 4.5) and extends from 1 to 10:

$$SMI = I + C$$

The San Marino scale is just a proposal so far, still to be formalized and adopted, but it can give the discussion on the dangers of interstellar transmissions a rational foundation and supply the basis for a regulation of the matter.

---

[22] I. Almár,"Quantifying Consequences Through Scales," *6th World Symposium about Space Exploration and Life in the Universe*, San Marino, March 2005.

Chapter 4

WHICH MESSAGE?

If and when we receive a message, the first important result will be the fact in itself: it will be proof of the existence of someone who broadcast it and therefore humankind will have the certainty that it is not alone in the Universe. This result would justify all the work scientists and technicians devote to the various SETI projects worldwide. Immediately after that, when the news will have caused the most diverse reactions from the public, the scientists who made the discovery will tackle an even harder job: understanding the meaning of the message.

The first possibility is that it is a transmission not directed to us or, in general, to a different species. We ourselves radiate transmissions of every kind into interstellar space; their power is enough to allow them to be received at a distance of many light-years with instruments similar to ours. If we think of a more advanced civilization, using more sophisticated instruments, our transmissions might be received at very great distances—provided, of course, that such advanced civilizations still use radio waves to communicate and listen to the Universe.

Because it is about half a century since we first broadcast these transmissions, by now our presence is known in a spherical space of a radius of about 50 light-years, containing more than a thousand stars. This number is much too low for a meaningful probability that one of them has a planet hosting a civilization technologically able to identify our transmissions, but this probability increases very quickly in time. Carl Sagan, in his novel *Contact*,[23] formulated the hypothesis that the first transmission an extraterrestrial civilization receives from Earth is the telecast of the opening of the Berlin Olympic Games of 1936 (70 years have passed, so that transmission has by now reached all the stars within a distance of 70 light-years, which involves several thousands). It is a fairly realistic guess, since that was the first high-power television broadcast. As Sagan noted, the first impression the extraterrestrials would have of us is that of a ceremony in perfect Nazi style, with Hitler in the foreground.

In chronological order, other very powerful transmissions were those broadcast by radar transmitters built by the United States and the Soviet Union during the Cold War, starting in the late 1940s. The only information extraterrestrials who received these transmissions would glean is evidence that we exist, together with some clues regarding our technology in the microwave field. However, these transmissions are directional and, owing to the combined motions of the antenna, the Earth,

[23] C. Sagan, *Contact*, Pocket Books, New York, 1985.

182

the Sun, and the planet on which the receiving antenna is located, they may be received for very short periods of time. If extraterrestrials follow the criterion of exceptional evidence, they will hardly draw any conclusion from this type of transmission, which constitutes a nonrepeatable event.

Today the most common and easily identifiable transmissions are television broadcasts, particularly those transmitted simultaneously by several stations. Sports events, like the Olympic Games or the Super Bowl of American football, are therefore our most likely interstellar presentations. Some very popular television serials are also a likely possibility, but there is also a nonzero probability that extraterrestials could discover our existence by receiving some of the porno videos that flood our planet, relayed by various satellites. In such a case our anatomy and reproductive modes will have no secret for their biologists!

These considerations suggest that it is very difficult to obtain a realistic picture of an alien civilization studying transmissions accidentally broadcast toward space, even when they may be decoded; to obtain images from a television signal without knowing anything about how it has been encoded (actually without even knowing that they are coded images) is no trivial task. And once we have learned how to transform the transmission into a message, the problem of understanding its meaning remains. In the past significant difficulties were encountered in interpreting inscriptions written in unknown languages. Etruscans belonged to our species, their civilization and their history are quite well known, and we have a lot of inscriptions in their language, yet our understanding of their writings is still limited. The decoding of an alien text, transmitted by beings about whom we know nothing and who are probably extremely different from us, is a perhaps impossible task.

If we ever receive a transmission addressed to us, or at least to a different species, it might contain a message coded in a simple way, with the aim of allowing easy decoding. There is one obvious problem: what do those who broadcast a message consider "easy" for those who are intended to receive it. The simplest alternative is to broadcast an image that must at any rate be encoded in a suitable way. The simplest way is to take a black and white image, subdivide it in a suitable number of points, and associate to every point the values 0 and 1, for instance, to represent a black or white point. The message of Figure 4.8 is a good example of a simple image encoded in this way. It is made by 73 sets of 23 binary digits, which can assume two values, 0 or 1, that is, white or black.

The idea at the root of this type of message is that anyone receiving many sets of signals, each one containing the same number of digits, is induced to put the various sets next to one another in such a way as to form a rectangle of white and black dots, forming a drawing. The limit to the resolution of the drawing is given by the number of dots in the message: at a rate of 10 bits per second to broadcast the 1679 (73 × 23) binary signals (1679 bits), less than three minutes are required. To broadcast a more detailed image, it is necessary to increase either the transmission rate, something that requires much more power, or the duration of the message, which increases the probability of an incomplete reception.

The total number of points (1679) must be the product of two prime numbers (in this case 23 and 73), so that there are few possibilities of reconstructing the image the wrong way. If, for instance, 1000 points were transmitted, a drawing with 20 lines of 50 points could be obtained, but also drawings with 25 lines of 40 points, and a large number of other possible combinations.

The message starts with the symbols for the numbers from 1 to 10. Then come the diagrams of the atoms needed for terrestrial life (hydrogen, carbon, nitrogen, oxygen, and phosphorus), followed by a representation of the double-helix structure of DNA. Under the DNA there is the sketch of a human figure and inside the double helix is the number 3,000,000 to suggest the complexity of human genoma. Next to the human figure, the number 4 billion shows the population of Earth at the time the message was transmitted. The message then carries a scheme of the Solar System, with the third planet under the human figure, to point out the origin of the message and a scheme of the antenna used for the transmission.

The very low number of bits used causes the resolution of the drawings to be poor. The human figure, for instance, is so schematic it allows different interpretations while the structure of DNA is so approximate it is quite difficult to understand. What an alien receiving this kind of message might understand is unclear. Interpretation would probably not be impossible if it is received by an alien similar to us, while in the case of a seriously different form of intelligent life, the result could be null. For instance, what might the sketches of the human figure and of the antenna mean to a being who does not have the sense of sight?

Perhaps by increasing the number of bits and therefore sending images with much higher resolution we could be reasonably sure that the message would be understood even by aliens rather different from us. After all, if they have an antenna able to receive radio signals, they must have a developed technology, and even if they have no sight it is unlikely that they could achieve such a technology without being aware of the shape of objects. They will perhaps combine the bits they receive to obtain a nonvisual representation of reality, which can be understood using their senses; in the end they should be able to identify the shapes.

But the greatest disadvantage of images is that they do not allow the communication of abstract concepts. Many authorities suggested sending messages based on mathematics, with the idea that mathematics is a sufficiently universal language to be comprehensible to any intelligent species, or at least to any species that has developed a technology. The universality of mathematics is not yet demonstrated, and the discussion on the nature of the truths it demonstrates is centuries old, but that's not the problem; it is unlikely that a language based on mathematics could allow the communication of concepts that go beyond mathematics itself, and rather elementary mathematics at that. It has been pointed out[24] that if a transmission without a real message (for instance, a nonmodulated carrier wave) would essentially say "here we are," a sequence of prime numbers or another mathematical message says "here we are . . . and we know some mathematics." This last statement adds precious little to the first one, since it is rather unlikely that someone who has a technology advanced enough to perform a transmission of this kind does not know any mathematics at all! If the intention is to send a meaningful transmission that goes beyond what can be said using images, it is necessary to use a language that allows for the communication of concepts—not an easy task in the context of SETI.

Toward the end of the 1950s, Hans Freudenthal, professor of mathematics at the University of Utrecht, elaborated a language that, in his intentions, would have been comprehensible to any intelligent being and called it *Lincos*, abbreviation for *Lingua Cosmica* (Cosmic Language).[25] *Lincos* is a language based on mathematics that allows any concept to be expressed in a perfectly logical way, at least in the intentions of its author. Actually the result is not so clear, as may be seen from the example reported below, taken from the book *The Science of Aliens* by C. Pickover.[26] The passage describes the generation of human beings:

---

[24] Neil Tennant, "The Decoding Problem: Do We Need to Search for Extraterrestrial Intelligence in Order to Search for Extraterrestrial Intelligence," *SPIE Proceedings*, 1967, pp. 50–59.

[25] Hans Freudenthal, *Lincos, Design of a Language for Cosmic Intercourse*, North Holland, Amsterdam, 1960.

[26] C. Pickover, *The Science of Aliens*, Basic Books, New York, 2000.

Ha Inq *Ha*:

$x \in$ Hom.$\rightarrow$Ini.xExt$\cdot$ $-$:Ini$\cdot$Cor$x$.Ext $=$
    Cca.Sec 11×10^{10111}:

$\wedge x \dot{} x \in$ Bes. $\wedge$:Ini.xExt $-$:Ini$\cdot$Cor$x$.Ext $>$Sec 0 :

$x \in$ Hom. $\rightarrow \dot{} \vee^{\lceil} y.z^{\rceil} : y \smallfrown z \in$ Hom. $\wedge \cdot y = $.Mat $x \cdot \wedge \cdot z = $.Pat$x$:

$\vee x \dot{} x \in$ Bes. $\wedge \dot{} \vee^{\lceil} y.z^{\rceil} : y \smallfrown z \in$ Bes. $\not\sim y = $.Mat $x \cdot \wedge \cdot z = $.Pat$x$:

$x \in$ Hom. $\rightarrow \dot{} \; \wedge t$:Ini$\cdot$Cor$x$.Ext:Ant:$t$:Ant:Ini$x$ Ext
    $\rightarrow$:t Cor$x$.Par$\cdot t$ Cor.Mat x :

$\vee x \dot{} x \subset$ Bes.$\wedge \cdot \wedge t$.Etc :

$x \in$ Hom.$\wedge$:$s = $ Ini$\cdot$Cor$x$.Ext$\cdot$
    $\rightarrow$:$\vee u.v^{\rceil} s$ Cor$u$.Par$\cdot s$ Cor.Mat$x$:
      $\wedge \dot{} $Pau Ant.$s \cdot$Cor v:Par:Pau Ant.$s \cdot$Cor.Pat$x$
      $\wedge$:s Cor x.Uni$\cdot s$ Cor$u$.s Cor$v$:

$\vee x$:$x \in$ Bes. $\wedge$.Etc :

Hom $=$ Hom Fem.$\cup$.Hom Msc :

Hom Fem$\cap$Hom Msc $= \lceil \rceil$:

Car:$\uparrow x \cdot$Nnc$x$Ext.$\wedge$.$x \subset$ Hom Fem
    Pau$> \dot{} $Car: $\uparrow x \cdot$Nnc$x$Ext.$\wedge$.$x \in$ Hom Msc :

$y = $ Mat$x$.$\wedge$.$y \subset$ Hom$\cdot \rightarrow$.$y \subset \cdot$Hom Fem:$\wedge$:

$y = $ Pat$x$.$\wedge$.$y \in$ Hom$\cdot \rightarrow$.$y \in \cdot$Hom Msc :

$x \subset$ Hom$\cup$Bes: $\rightarrow$:Fin.Cor$x \cdot$ Pst.Fin$x$#

Literally, the text says:

The existence of a human body starts some time before that of the same
human being. The same occurs for some animals. Mat, mother. Pat, father.
Before the individual existence of a human being, its body is part of the body
of its mother. It is originated from a part of the body of his mother and from a
part of the body of his father.

In truth, the problem of language has probably been greatly overestimated. The difficulty in decoding a message depends much more on
the possibility of associating meanings to the symbols than on the structure
or difficulty of the language. Paradoxically, the radioastronomer Jean
Heidmann proposed broadcasting the *Encyclopedia Britannica* directly.

What has been said concerns the decoding by extraterrestrial intelligences
of a possible message from Earth, but the problem would be similar in the
case of a message *we* might receive. Even if we succeeded in decoding the
message perfectly, an infinity of questions would remain. First, up to which
point should we believe in it? The most complex messages sent by
humankind into space, the disks carried by the *Voyager* probes, will be
described in Chapter 5. Many say the description of humans they convey is

not what humans are but how they want to appear, maybe more to themselves than to others. In preparing the message a kind of self-censorship was exercised to exclude all images of violence, those that could appear indecent (causing difficulties in describing human anatomy in a comprehensible way), and those that could be interpreted by a human group as an abuse of power by another group (no image describing anything linked with religion was included). Apart from the obvious case of a message aimed at fooling other species, how can we distinguish their description of reality from fiction, or how can we interpret their art? What if we receive their equivalent of *Star Trek* or a horror movie?

The search for extraterrestrial intelligence is performed mainly using radio receivers, since it is a common assumption that radio waves, particularly at frequencies of about one gigahertz, are the simplest means to send messages over large distances. As already pointed out, this common opinion could be an error of perspective due to the present state of our technology. Some scientists started a search in the optical zone of the spectrum (so-called optical SETI), looking for messages broadcast using devices that might be similar to our lasers, signs of the presence of extraterrestrial intelligence, like large engineering works on a cosmic scale or flashes of light emitted by hypothetical spaceships.

As will be seen in Chapter 5, there are infinite ways to send messages using material supports. The engraved plates carried by the *Pioneer* probes or the phonographic disks on the *Voyagers* belong to this type of communication, but at our level of technology we can already imagine hundreds of other ways to record a message, from holograms to living beings. Information might be coded in the DNA of a living being, since in every organism there are long chains of DNA that are not used for storing useful information and could therefore be manipulated and used for other purposes, such as recording messages. If hypothetical aliens used complex means to code information on some support, it is likely that we would not even realize it, and then the probability of our succeeding in decoding the message would be extremely low.

EXTRATERRESTRIALS, HOW?

The difficult choice of broadcasting messages or not might be solved if only we knew the possible recipients, but despite the endless number of hypotheses that have been advanced, we have no idea of what aliens might be like.

A basic assumption of modern science is that the laws of physics are the same throughout the Universe. Does this uniformity allow us to think that the bases on which life may develop are also uniform? If biology follows universal laws and the processes of evolution and adaptation to the environment are everywhere the same, then the biochemistry of life—based on chains of amino acids and on proteins—in other biospheres will not differ much from that on our planet. This obviously does not mean that beings that evolved in more or less similar environments will be similar, but simply that they will have common features, at least at a basic level.

Yet nobody can exclude the existence of forms of life that may be based on very different laws or a different biochemistry. Beings based on silicon and sulfur instead of carbon and oxygen have often been imagined; it is true that the variety of the compounds of silicon is much smaller than the compounds of carbon, but we do not know whether this is enough to state that life based on the former has little chance to start and evolve.

The discovery of extremophile beings on Earth should make us think about the possible variety of life-forms. Yet it is possible that such variety is reduced when evolution moves from bacteria to higher life and that only particular types of life can progress to actually give way to animal life and then to intelligence. If consciousness and intelligence are due to complexity, it is predictable that only a few of the various forms of life will reach a sufficient complexity.

Even without considering extremophiles, living beings on Earth evolved to a truly amazing variety of shapes and sizes, and even more impressive is the variety of behaviors and reactions of the various species. Yet all descended from a common ancestor and evolved in environments strongly similar to each other, at least if compared with the environments that can be found on other planets. Their DNA has the same structure and often, in many parts, an amazing similarity. Many animals that are not closely related to us (in the sense that they belong to other phyla), but have evolved in our same environment and now live close to us, are so strange (that is, different from us and from animals that can be considered to be our relatives) as to be much more "alien" than most aliens in science fiction. The simplest example are insects: if they were not so small and we could see them in detail, they would appear as unimaginable monsters. The appearance of extraterrestrial beings is therefore probably such that we cannot, with all our imagination, picture it. But it is not only the physical aspect: the behavior of insects, for instance, is so different from that of the animals nearest to us as to be even stranger and more unintelligible than their appearance. The first example that comes to mind is that of the social insects, such as bees or termites, which seem to display a kind of rudimentary collective intelligence, not accompanied by either intelligence or consciousness at the individual level.[27]

But the behavior of other invertebrates is hardly less strange, particularly in the field of reproduction: from the insects that lay their eggs inside other insects of a different species so that a newborn can feed on their guest once the eggs open (this behavior was probably chosen by the scriptwriters of *Alien* as a model for alien monstrosity) to others in which newborns, as soon as they are out of the eggs, suck their mother's vital juices to the point of eating its body completely, or to the many cases in which the female not only eats the male after (or in some cases during) intercourse but males almost seem to let themselves be eaten voluntarily (note how anthropomorphic this last sentence is, and above all the use of the word "voluntarily"). If the variety of forms and behavior on Earth is so great, what might we expect from extraterrestrials?

Yet some assert that this variety must decrease as we rise along the evolutionary scale and that intelligent beings must be more similar to each other than we might expect from the considerations above. The phenomenon that goes under the name of convergent evolution is often mentioned: completely different species, belonging to phylogenetic lines quite far from each other, developed incredibly similar organs under the pressure of similar evolutionary conditions. If, for instance, sight has an important value for survival, at least in an environment rich in electromagnetic radiations in the visible region of the spectrum (or better, in the region that we on Earth refer to as visible), and if to detect such radiations a lens focusing them on a sensorial surface is needed, it is likely that beings in separate biospheres will develop eyes not too dissimilar from each other. If mechanisms of this type are important in determining the nature of living beings, the problem regarding their variety is restricted to that related to the variety of environments in which complex forms of life may develop.

One of the problems much discussed in the past is the so-called *predominance of the humanoids*, that is whether intelligent living beings must have an appearance similar, in a general sense, to that of humans. By humanoid we mean here a living being with bilateral symmetry, endowed with two legs and two arms, with organs for manipulating objects, with the brain located in a head above the bust, where the main sensory organs

---

[27] Bernard Le Bovier de Fontenelle, in his book *Conversations on the Plurality of Worlds*, published in 1686, uses the example of social insects to show the odd behavior of many beings living on our planet and concludes that the beings living on other worlds are likely to be so strange that we cannot even imagine them.

are also located, and in particular two eyes allowing binocular vision, together with an opening to introduce food.

Supporters of this thesis start from the idea that the humanoid configuration is particularly suited to a highly intelligent being. The control of a biped's equilibrium is complex and requires a well-developed nervous system and brain. If a single "control center" is present, or better, if evolution brings increasing encephalization, as happened on Earth, biped stance and intelligence strengthen each other: the requirements of equilibrium control causes an increase in the mass of the brain and the increase in brain size facilitates the biped stance.

Evolution on Earth is characterized by a gradual reduction of the number of legs: from the filaments (parapods) of annelida (for instance, the millipedes) to the articulated legs of the arthropods, a constant reduction in their number (10 in crustacea, 8 in arachnida, 6 in insects) occurred. With terrestrial vertebrates, the number of legs was reduced to four. A high number of legs, together with a low position of the center of mass (that is a small height of the center of mass if compared with the "track" of the legs) allows the animal to easily remain in static equilibrium during all phases of walking. A quadruped, particularly if its center of mass is not that low, must, in each step, go through positions that are not well balanced and must therefore coordinate movements with greater precision and have quicker reactions than a hexapod or an octopod. Besides, the larger the animal and the lower the gravity of the planet, the easier it is for it to remain in equilibrium on fewer legs, in the sense that the response of the nervous system to avoid falling down need be less quick. From this point of view, low gravity simplifies the operations linked with motion.

On the other hand, a maximum walking speed exists for any animal; to go faster the animal must change its gait and perform a transition from walking to running or jumping. This speed is faster for taller animals and higher gravity. Since in general high speed is an important factor in natural selection—whether to run away, to chase food, or to anticipate others in the search for it—there is a strong incentive to shift from walking (a sequel of static balanced positions) to running (an alternating of balanced positions and others in which equilibrium is not guaranteed). Large animals, possibly with a smaller number of legs, may then have an advantage. Very large animals, not able to use a biped stance, may use the tail as a support for stabilizing the body while walking on their hind legs: the tyrannosaurus is an example of an animal that adopted an erect posture thanks to this third support, even if it surely had a primitive brain.

This reasoning does not prove that all intelligent species must be bipeds, but it is a good clue in this direction and it would be surprising to find an intelligent gastropod or an intelligent being with a lot of legs. It is likely that the erect position is linked with the biped stance and that the position of the head above the trunk and not forward is made necessary by the increase of the brain's weight. The setting of the brain inside a strong bone box and its position close to the organs of hearing and sight seems reasonable, as does its location as high as possible to obtain a better view of the surrounding environment. The position of the head above a more or less articulated neck allows the environment to be kept under better control and partly obviates the drawbacks linked with the frontal position of the eyes, a characteristic allowing binocular vision, which is important to evaluate distances. Obviously, a larger number of eyes could be even more advantageous, but perhaps it would require a significant increase in the volume of the brain: a nonnegligible part of the brain of higher animals is devoted to processing signals from the eyes. All complex terrestrial animals have two eyes but very few have more: perhaps it would be a waste to devote precious "computing power" to sight, over that needed to elaborate images supplied by two eyes. And by the time a brain large enough to afford the "luxury" of more eyes is available, the general plan of the body is consolidated and it is too late to evolve a different configuration.

A position of the arms that brings the hands within the field covered by the eyes is certainly of advantage; therefore, the humanoid configuration also seems to be a good layout from this point of view. Certainly a larger number of arms would be an even better solution, as those who are accustomed to manual work know all too well, but perhaps it is unlikely that a configuration of this kind may evolve. The arms are an evolution of the front legs, and if evolution favors quadrupeds over hexapods, it is not likely for other auxiliary limbs to develop subsequently, once the configuration with four legs has been fixed through millions of years of evolution.

A hypothetical example of convergent evolution and of *predominance of the humanoid form* is the result of the study by paleontologist Dale Russel of the Canadian Museum of Nature. Starting from the hypothesis that the catastrophe that wiped out the dinosaurs, and indirectly delivered Earth to mammals, had never happened, he simulated the evolution of reptiles toward intelligence. The result is a humanoid 137 cm tall, called *dynoman* by researchers (Figure 4.9); the similarity of the body structure with that of *Homo sapiens* is apparent.

If the humanoid shape is probably optimal for an intelligent land animal, it is not suited to aquatic life or to flying. But is it possible for an aquatic or flying intelligent species to evolve? Many authorities deny this possibility.

FIGURE 4.9 Model of a hypothetical humanoid deriving from the evolution of dinosaurs, next to his ancestor, the stenonychosaurus, a small dinosaur living from 70 to 80 million years ago in Canada. An enlargement of the head is also shown (from T. Dickson and A. Shaller, Extraterrestrials – A Field Guide for Earthlings, Camden House, Camden East, Ontario, 1994).

Aquatic life is generally not that favorable to the development of intelligence and all aquatic animals, with the exception of cetaceans, show very little intelligence. Cetaceans evolved on dry land and only later adapted to water life, bringing to their new environment a level of intelligence they acquired elsewhere. At any rate dolphins, despite their big brain, did not progress very far along the road to intelligence, probably because of the impossibility of manipulating objects. Besides, a very important stage in human evolution was the use of fire, something not possible for aquatic species. Perhaps dolphins are a living example of the impossibility of true intelligence without technology.

The shape of large sea animals probably follows two fundamental types: that of dolphins and that of rays. The first is the most suited to fast swimming in deep waters, and the fact that many fish of different families (like sharks and tunas), sea reptiles (ichthyosaurus), and cetaceans (dolphins) developed almost identical shapes despite having evolved in completely independent ways has been considered by many as evidence of the power of convergent evolution. The shape of the rays is optimal for motion close to the seabed, or for motion in a fluid close to a solid surface, as studies on the aerodynamics of motor vehicles showed. In both cases they do not look like shapes that are very suitable for an intelligent being.

In a similar way, many consider it unlikely that a flying species may develop intelligence, since the weight of a large brain is certainly an obstacle to flight; besides, controlling flight is probably simpler than it seems at first sight, since the number of neurons devoted to this function in many insects is quite low. However, there is no doubt that good control of the shape of the wing to obtain greater aerodynamic efficiency is helped by the presence of a powerful brain, which is also useful in performing complex maneuvers like takeoff from very rough ground. The drawback of the brain's weight might be easily overcome on planets with a low gravity and a dense atmosphere, where a flying intelligent species might exist.

Humans, like most Earth animals, get data from their environment through five senses. It could be said that *seeing* consists of detecting electromagnetic waves in a certain range, *hearing* consists of detecting pressure waves in the surrounding medium, *smell* consists of detecting the chemical nature of what is carried in the medium and enters the respiratory system, *taste* consists of detecting the chemical nature of what enters the digestive system, and *touch* consists of obtaining information about what enters in contact with the body. While the frequency range of the electromagnetic or pressure waves, or the sensitivity of the sensory organs, may differ widely from one environment to another (or, as terrestrial biology shows, even between organisms that evolved in the same environment), it is difficult to imagine a radically different interaction with the environment. As for touch, manipulatory organs must have some provision to evaluate the force exerted and the characteristics of the surface on which they operate. Similarly, a set of proprioceptive organs must be present to evaluate the relative positions of the various parts of the body.

The senses listed above must be interpreted in the widest possible way: the organs that might enable a being living close to a gamma-ray source emitting very little light to detect gamma-rays will be referred to as sight organs. Similarly, organs able to detect pressure waves in a fluid will be referred to as hearing organs, even if they have little to do with sound. With these broad definitions, it is impossible for us to imagine other senses that might add to ours or substitute for them. We must then conclude that either they do not exist or we are so blind to a part of the real world that we miss some of its features completely. This is the thesis of those who support extrasensorial perception (e.g., ETIs with telepathic powers, etc.), but there is no proof, or even a serious clue, that these things exist at all. In conclusion, we can assume that any extraterrestrial being has the same senses we have, although some may be much more or much less developed; for example, seeing organs can just detect light intensity.

Hearing might be so developed that it affords a whole picture of the outside world, instead of just detecting a few properties of the pressure waves, as our ears do.

Finally, what is the best size for intelligent beings? Probably a minimum size exists, since the size of organic molecules is the same in the whole Universe and living structures, at least if they are based on the biochemistry we know, cannot be too small. If to reach intelligence a minimum of complexity is needed, it is likely that intelligent brains of too small a size and mass cannot exist. The same should also hold for different configurations, for instance a nonencephalized intelligence, that is, one based on a nervous system distributed in various parts of the body, though we do not know whether a configuration of this kind is at all possible.

The smallest dimensions of an intelligent being are therefore those of a body able to support, move, and feed a brain of sufficient size. The maximum size is probably linked to the environment and in particular to gravity. In general, large dimensions can be an advantage, since a big animal can be faster (at least if it is not too big), can defend itself better from predators, see farther, and so on. But they are also a drawback, since a big animal needs more energy and must therefore eat a larger quantity of food, something that, on the one hand, might help the development of intelligence, since it creates problems that have to be solved, but, on the other, also leaves less time available for activities not directly related to the mere search for food. And beyond certain limits, food may not be available at all. As a general rule, we might expect intelligent beings that evolved on planets with lower gravity to be larger, and it would be surprising to find very small extraterrestrial intelligent beings, as small, for instance, as insects or perhaps even as a mouse or a rabbit.

The above reasoning leads us to think that the humanoid form, in a very general sense, with a size not too different from that of humans (let's say from about a quarter to four times as big), is the most probable layout for an intelligent extraterrestrial. A group of humanoids warming themselves by a fire is shown in Figure 4.10. The fire should show that they are intelligent beings, but the fact that no tool is represented (and that they are naked, in spite of the cold suggested by the presence of the fire) implies that they are indeed very primitive. The scenario described above is thought by most researchers today to be much too anthropomorphic and much too "Hollywood-style." The look of the species in Figure 4.10 reminds one of popular representations of aliens, and particularly of the descriptions given by persons who claim to have been abducted by them. On this and on the similarity of these aliens with human fetuses, see Chapter 5.

FIGURE 4.10 *Sketch of a hypothetical group of aliens warming themselves around a fire. The fire suggests that they are intelligent, but is it possible that intelligent beings have nothing with which to cover their skin when it is cold? (From T. Dickson and S. Shaller,* Extraterrestrials – A Field Guide for Earthlings, *Camden House, Camden East, Ontario, 1994.)*

Contrarily to everything discussed above, many today think that the humanoid form does not prevail. First of all, the very concept of convergent evolution is contested. For example, it is true that the ichthyosaurus, the shark, and the dolphin developed very similar body shapes, but to claim that these life-forms are distant from one another in evolution is only an error of perspective. Fish, reptiles, and mammalians all belong to the chordate phylum and are genetically very closely related. Convergent evolution, therefore, would be possible only if the form under examination is already somehow contained, at least potentially, in the genes of a common ancestor, in a play of divergence–convergence that at any rate requires not too weak a relationship. So evolutionary convergence between living beings that do not have anything in common, like those that evolved on different planets in an independent way, must be excluded.

Things are different when dealing with the development of a certain organ for a certain function. In this case it is possible that the anatomy is so dictated by the physical principles of operation that the result is similar configurations, even if the living beings belong to completely independent evolutionary lines. An example is that of the eyes. To only detect the presence of light and perhaps its intensity, a photosensitive zone of the skin is enough. But if a real image is required, a lens focusing light onto a photosensitive zone is needed, and the latter must be connected with the brain through a number of transmission lines. The eye was developed several times independently, but we must note that the results have not been so convergent as it seems at first sight (excuse the pun): eyes of various types exist and the differences between, for instance, the multiple eyes of many insects and those of vertebrates are large. The common feature of all eyes is that they are organs of sight, yet the vision they allow may be very different from case to case. Not only can the field of frequencies be rather different (rattlesnakes see in the infrared range, bees see ultraviolet light), some animals see colors, others do not; some have a system of vision particularly good at seeing moving objects, etc. And all these different devices have been developed by beings that have many common genetic characteristics.

Even taking into account convergent evolution, it is clear that different environments cause strong differences in the case of intelligent beings, too. We may wonder if some environments are more suited to the development of intelligence, while others are not suited at all, as for instance the sea might be.

Carl Sagan once suggested that living beings with balloons full of gas, allowing them to float and never sink into the denser layers, may have evolved in the atmosphere of giant planets. A hypothetical example of extraterrestrial intelligence suited to giant planets is shown in Figure 4.11: a huge being, 500 m long, living in the upper layers of the atmosphere, floating like a balloon, living in the aerostatic cities represented in the background. If beings of this kind exist, they might be more common than we, the inhabitants of rocky planets, are; we have evidence of the existence of many giant extrasolar planets, but not of many planets of terrestrial type.

But is this hypothesis realistic? Probably not: the being in the figure has no hands with which to manipulate objects, which by the way could be quite hard to get in a world made only of gas and perhaps droplets of liquid. With what could he build his cities? And, if intelligence is accompanied by technology, how could he develop it? The inhabitants of the giant planets could perhaps reach the level of intelligence of dolphins or other cetaceans.

Humanoids have a bilateral symmetry, that is they have a plane of symmetry at least for their external shape, while the organs inside the body may be asymmetrical (the liver on the right, the heart on the left, etc.).

FIGURE 4.11 *A hypothetical intelligent being living in the upper layers of the atmosphere of a giant planet like Jupiter. The giant aerostatic cities in which these enormous intelligent balloons live are represented in the background (from T. Dickson and A. Shaller,* Extraterrestrials – a Field Guide for Earthlings, *Camden House, Camden East, Ontario, 1994).*

This type of plan is typical of most animals along terrestrial evolutionary lines, from arthropods upward, and it is so natural for us that whoever drew the being in Figure 4.11 intending to describe something really alien followed this layout. Yet bilateral symmetry is not universal in the animal kingdom on Earth and was not immediately adopted; coelenterates exhibit a radial symmetry, like many other sea animals, from starfish to octopuses.

All animals with a bilateral symmetry have an even number of legs and the few that stand on three supports use a strong tail for this purpose. Likewise they have an even number of organs for sight and hearing and, if they have a single mouth and a single anus in the symmetry plane, this is due to the fact that they have a single digestive tube. Yet there are two nostrils, so the single nose has two openings. Beings with a radial symmetry may have any number of legs (or better, of tentacles, since no being walking on legs on land has a symmetry of this type), and an odd number of sensory organs.

There does not seem to be a definite reason for animals living on dry land to have a bilateral symmetry, if not for the fact that they descend from ancestors that had this configuration. If, on some planet, animals with radial symmetry developed a skeleton and a set of articulated legs and lived on a solid surface, perhaps it is possible that an intelligent species evolved from them. It is possible to imagine configurations based on radial symmetry as suitable for hosting intelligence as the humanoid one, for instance with base three (three legs, three eyes, etc.). But other possible symmetries exist, apart obviously from the absence of symmetry; we must not forget that the simplest beings, like the amoebas, have no defined shape and therefore no symmetry.

The debate on the prevalence of humanoids is still open, the solution being perhaps not that intelligence develops mainly in humanoid species, but that the planets on which humanoids can develop are those in which it is more likely that evolution will lead to intelligence. Besides, it is necessary to consider that if, among the forms that evolved on Earth, humanoids are particularly suited to intelligence, other forms never seen on this planet and suited to hosting intelligence might exist. Evolution proceeds at random and, when it finds a suitable form, works on it through small variations: it certainly does not try all possible solutions. As mentioned earlier, it is likely that if intelligent species exist, they are much older than ourselves. Might it be that, even if intelligence developed as a humanoid species, it will eventually evolve into something different? Here, too, there are many questions to which we do not yet have the answers.

Some hold that physical evolution came to a halt with the birth of intelligence and cultural evolution took its place. This statement seems arbitrary, but is probably more justified than it sounds. Thanks to intelligence, humans interfere heavily with the mechanisms that promote evolution. The simple fact that humans value human life so much leads to opposing natural selection with increasing success as soon as science and technology advance. Medicine is after all nothing other than a long battle against natural selection: humans do not want survival of the fittest to apply to their species, and strive to grant the longest possible life to each individual, even those least fit to survive. This undoubtedly favors the survival of mutations that, being disadvantageous, would disappear quickly in a regime of natural selection.

Medicine, therefore, must also face the problem of mending the damage to the human genome caused by this attitude. The knowledge of the human genetic code and of the techniques needed to modify it, correcting the defects carried by the individuals who would not survive in a natural situation but do so only by the use of medicine, are essential for this task. In this way an intelligent species works to maintain its own genetic inheritance unchanged, stopping the mechanisms (mutation and natural selection) of evolution.

There are no doubts about the ethical correctness of this way of proceeding and it is a precise duty of medicine to grant a life as healthy and normal as possible to all individuals. But today, and probably more so in the future, medicine could also have another task: not to stop evolution but to facilitate it by producing favorable mutations. The current consensus on this subject is that practices of this type are morally condemnable. The negative moral judgment, however, may be motivated by two orders of reason: the intrinsically illicit character of any practice directed at modifying the human species, and the wish to avoid any intervention for fear of causing damage instead of bringing benefits. In the first case the ban on these practices must remain forever, while in the second it is only provisional; when the techniques in this field have improved, what is forbidden at present might become legitimate.

There is no intent to enter here into the bioethical debate, but simply to say that, whatever the ethical approach of humankind, a different intelligent species might see it differently and species might exist that have taken their biological future into their own hands: they might have entered a phase of rapid artificial evolution. As noted by ethnologist Ben Finney, the future evolution of humankind is linked with its expansion in space.[28] The communities that settle other planets will necessarily be small, relatively isolated, and will live in different environments from that of Earth, and therefore evolutionary pressures will be strong. Moreover, these environments might easily be rich in radiation, which will facilitate mutations. Besides, the push to modify humans to adapt to the local environment might become strong enough to overcome, in some cases, moral restraints.

No doubt this will be true for people living in large space habitats. It may begin with simple modifications aimed at reducing the physiological consequences of exposure to low pressures and the absence of gravity. Certainly a genetic intervention that reduces the consequences for the workers involved in an accident that implicates partial depressurizing, or

---

[28] B. Finney, *From Sea to Space*, The Macmillan Brown Lectures, Massey University, Hawaii Maritime Center, 1992.

permits those who must spend a long time in low gravity (for example, miners working on asteroids) to avoid long periods of adaptation and rehabilitation, does not raise particular ethical problems. Since the cost of construction and maintenance of a habitat is lower if the need to pressurize and generate artificial gravity (through rotation of the whole space station) is less severe, these interventions may lead to better adaptation of humans to space conditions, with the creation of a new human race and then, eventually, of a separate species. These are very long processes, but we must remember that the times considered in this context are of millions and millions of years. In the long run there might well be a differentiation between humans living in space and humans living on planets and, among the latter, even differences among those living on the various planets.

These considerations apply to all intelligent species that are spread in space and it is therefore quite possible that the alien species with which we will come in contact will really be groups of species, more or less differentiated within themselves.

The absence of further morphological changes in humans since *Homo sapiens sapiens* appeared raises the question of the apparent end of evolution once intelligence has been reached. Two possible answers, as outlined earlier, are a shift of evolution from the biological field to the cultural one and a continuation of biological evolution, possibly driven by genetic engineering. But there is a third possibility, particularly favored by some artificial intelligence experts: a shift of the evolutionary line from carbon biochemistry to inorganic chemistry based on silicon, with the construction of intelligent machines able, with time, to replace humankind.

Today this sinister scenario is (happily) losing strength, but the alternatives are not much more reassuring. In fact, though many artificial intelligence experts continue to think it possible to build intelligent computers starting from the architecture of the present machines, the difficulties in making progress in this field suggest that the problem may not be solved with more powerful hardware and suitable software. The opinion is widespread today, as mentioned when speaking of Von Neumann machines and probes, that true artificial intelligence is not achievable using conventional computers. A possible way to build intelligent machines, pursued by not a few researchers, might be by so-called quantum computers. If, with the help of nanotechnologies and quantum computers, intelligent machines can in fact be built in the future, the perspective will not be substantially different from that of classical artificial intelligence; the type of hardware changes radically but the idea that the next step in evolution is the construction of machines that can take the place of living beings is essentially the same.

This is an alternative suggested by those who think that the human brain is nothing more than a complex biological computer, and the human mind simply the software running on it. From this view, meant in a literal and not metaphorical sense, derives the possibility of displacing a human mind to a machine with sufficient computational power, in a way not much different from what we normally do when transferring a program from one computer to another. If human beings "in their present implementation" have all the limits of living beings, including mortality, transferred to a new hardware, they might transcend them. These are just arbitrary speculations, and nobody has ever explained how it might be possible to perform such a transfer. We can hope that these perspectives remain at the level of ideas, and are only applied in a science fiction (or, rather, a horror movie) screenplay.

Progress in bioengineering and nanotechnologies will certainly make it possible to build machines containing biological components and to replace biological organs with mechanical prostheses in living beings. Science-fiction writers often imagined partly biological and partly mechanical beings (the so-called cyborgs), and the disturbing prospect that they might represent the next step in evolution cannot be avoided. We therefore must not exclude the possibility that the extraterrestrials with which we will come in contact will not be living beings in the literal sense of the term but intelligent machines "bred" by an intelligent species that has then been extinct for millions of years. This could be of little importance if the contact were made by radio; an alien artificial intelligence probably will not differ substantially from that of its builders, and perhaps we would never notice any difference. Even in the case of a dialogue, we may converse with some machine and think we are speaking with living beings. If direct contact can be established, however, things could be radically different, but this aspect will be dealt with in Chapter 5.

We came to the problem of the predominance of humanoids among intelligent life-forms from a phenomenon that has been called "convergent evolution." The possibility that convergent evolution could actually occur between species that evolved on different planets is much greater if life passed from one planetary system to another in the form of microorganisms with their sets of genes. Yet someone noted that convergent evolution is a general phenomenon, not restricted to the biological field. Cosmological and chemical evolution also show converging aspects.

These are controversial statements, but there are scientists who think that evolution at the cosmic level is not entirely dependent on chance but is, at least up to a certain level, predictable. Likewise they think that chance has a limited role in key steps at all levels of evolution, and that characteristics like the cellular structure of living matter, eukaryogenesis, multicellularity, neurogenesis, and even the development of intelligence are to a certain extent obligated stages, made necessary by the uniformity of the fundamental laws of physics. The uniformity in the structure of stars and galaxies gives a convincing experimental basis to convergent evolution in the cosmological field, and the presence of the same organic compounds in the Solar System and in interstellar space is evidence in favor of the convergence of chemical evolution, at least in our zone of the galaxy and for the simplest organic molecules.

Does convergent evolution play a role in determining the nature of intelligence? If that were the case, intelligent species of independent origin might have at least some features in common, while in the opposite case it could be difficult even just to recognize each other as intelligent species. In the latter case, the worldview of intelligent species may be so completely different and incommunicable that any form of exchange between them may be impossible. On the contrary, Marvin Minsky asserted that the common limitations of time and space and the universality of natural laws cause the epistemological horizon of all intelligent species to coincide, without the need to resort to convergent evolution. It is clear that between these extreme positions there is an infinity of intermediate approaches, depending on which different intelligences may reach a more or less partial agreement.

So far it was implicitly assumed that intelligence (and thus technology) and consciousness are two different aspects of the same phenomenon, linked in some measure to the increase in cerebral mass or, better, to a more complex structure of the brain. If consciousness is in fact tightly linked with complexity, this formulation is correct and a high degree of consciousness cannot be separated from intelligence. If, on the other hand, consciousness is not a direct consequence of complexity, but developed as an answer to evolutionary pressures that do not necessarily coincide with those leading to intelligence, it is not impossible that beings could exist that are endowed with just one of the two.

Leakey[29] and other paleontologists think that consciousness developed as a response to the need to deal with the dynamics of personal relationships within primitive societies, first of primates and then of hominids. The complex struggle for supremacy—directly linked to the possibility of having many descendants, and therefore allowing the genetic

[29] R. Leakey, *The Origin of Humankind*, Orion Books Ltd., London, 2000.

inheritance to spread—makes the ability to foresee the behavior of other members of the society a very important feature. Consciousness of the self and of one's own motivations would allow us to rationalize our knowledge of the actions of others and to play a winning role in the struggle for power. An analogy can be made with the game of chess: a computer plays by calculating an enormous number of the opponent's moves and countermoves and choosing, mechanically, the strategy that seems to be more favorable, without the intervention of intelligence or consciousness. The human player, who has neither the memory capacity nor the computing power of the machine, tries .to understand his opponent and to foresee his moves; to do this, both intelligence and consciousness are essential. In chess it has been shown that by now the brute computing power of a machine is able to overcome the reasoning of conscious minds.

This explanation of consciousness has an interesting consequence: if it is an instrument to understand the behavior of others, we must assume that others use the same mechanism we use to understand the world and to react to it. It is only logical that we extend this approach, even if arbitrarily, to other living beings and also to inanimate things, personalizing animals and natural phenomena. This may be the origin of our tendency to anthropomorphize, and there is no doubt that we will tend to apply it to aliens, when we discover them.

If it were true that this is the origin of consciousness, we could imagine beings that have intelligence and technology without being conscious; in some extraterrestrial environment, where conditions were favorable, it is not impossible that intelligence without consciousness could develop. Likely, the technological progress of intelligent beings of this type would be very slow, exhibiting the characteristic time spans of biological evolution.

An intelligence of this kind is almost unimaginable for us and perhaps it is even more difficult to imagine a conscious being that has not developed intelligence. Yet the fact that we cannot imagine such beings does not mean that they cannot exist; as pointed out more than once, this is what we must expect from beings who evolved in a completely different way from us. Then there is another possibility that beings may be, or at least appear to be, intelligent despite not being conscious; if indeed other technological civilizations developed machines that, even if not really intelligent, are able to control, maintain, and perhaps also replicate themselves in an autonomous way, these machines may well survive their builders. We could then come into contact with machines that continue to emulate the intelligent behavior of their creators, without being intelligent or conscious. In this case it might be extremely difficult to realize that we are not facing a true intelligence; it would be a sort of Turing test, but with an intelligence of unknown type and with very long times between transmitting the messages and receiving the answers.

Evolutionary pressure may have played an important role in determining not only the form of the body and the structure of intelligence, but also the behavioral rules and the values that are the basis of how we define our human nature. Tolerance, loyalty—at least within the group—and helpfulness toward other people bring advantages, at least in the long run, for the person who exercises them and they may have been favored by the process of natural selection. Convergent evolution may have played a role in determining a common base of values on which intelligences of different origin can agree.

Among researchers who operate in the SETI field there is widespread consensus that any intelligent species with which we may enter into contact will be much older than us. This consideration derives from the fact that our species is less than two million years old, but has been able to communicate at interstellar distances for less than 50 years. The problem is that if the duration of an intelligent species (the L factor in Drake's equation) is short, the probability that two intelligent species exist at the same time in our galaxy and are able to communicate with each other is extremely low. Only if intelligent species are long-lasting or if they maintain their ability to communicate for very long times, in the order of millions of years, will the probability of contact be high.

If, for instance, the average duration of an intelligent species is 100 million years (still short, on the time scale of cosmic phenomena) and if the ages of the species now extant are evenly distributed, the probability of contacting a species as old as us or younger is 1/2,000,000 (that is, if we contacted two million species, only one would have an age similar to ours) and the average age of the civilizations with which we may have a relationship would be 50 million years. These are only very rough estimates, but they confirm that we will either never succeed in contacting another intelligent species, or the contact will happen with a species more ancient than ours.

Other researchers, on the contrary, think that we could be one of the oldest intelligent species in our galaxy. Among the possible causes of mass extinctions are the already-mentioned "gamma-ray bursters" that have been observed in distant galaxies. In the last few years more than 1000 such events were recorded—a sign that they are not too rare. A gamma-ray event in our galaxy could have canceled most of the species with the exception—thanks to the shielding effect of water or rock—of those living in the depths of the sea or underground. If the event lasts for a short time in comparison to the period of rotation of the planet, it is possible that it could affect only a part of its surface (in any case more than a hemisphere), in which case the number of species annihilated will depend on chance; if, in the last million years, Earth had been struck in the hemisphere containing Eurasia and Africa, humankind (or its forerunners) would have been completely eliminated. Events of long duration could have even more extensive consequences.

Some scientists ascribe the mass extinction that extinguished the trilobites to a gamma-ray burster. The effects of these explosions on the development of intelligent species could be quite strong, particularly if the development of intelligence has a low probability of occurrence; in this case, if a species on its way toward intelligence is eliminated, millions and millions of years could pass before another species starts the same development. It seems that gamma-ray events are less and less frequent with time; if this is true, the development of intelligent species might be something recent, and planets much older than Earth would not have hosted higher life-forms for long because of the frequent mass extinctions. Earth itself would have had to wait a long time before hosting sufficiently complex animals. Perhaps this argument does not allow us to conclude that intelligent species older than us by hundreds of thousands or some millions of years do not exist, but, if its premises are correct, it could rule out that intelligent species developed hundreds of millions or even billions of years ago. Yet this is just a hypothesis for now, whereas the opinion that if extraterrestrial intelligent species exist they must be much older than we are, is very well rooted.

Such ancient intelligent species are absolutely unimaginable for us, just as it is impossible to imagine what we might become millions and millions of years from now. Yet there is a general agreement that such old species must necessarily be more advanced than ourselves, not only in a technological sense but also where ethics are concerned. In short, they would be extremely wise species that succeeded in overcoming their own aggressiveness (not least because they would not survive for so long otherwise) and would teach us not only scientific truths and technological know-how, but also moral truths, playing the role of spiritual guides.

This vision may be just another aspect of the tendency to anthropomorphism and reasoning by analogies that are so common when speaking of extraterrestrials.

An argument advanced by Frank Drake to support this thesis is based on the continuous extension of the average life span of our species and on the hope that this trend may continue thanks to bioengineering and the knowledge of the human genome. It is reasonable to assume that any intelligent species whose body is biochemical, like ours, will operate in a similar way and that an extremely ancient species will therefore be much more long-lived. Individual longevity also answers the problem of the deceleration in cultural and scientific–technological evolution that seems necessary so that a species can indeed last for times on the order of millions of years; the slow-down in generational renewal will surely result in a much less dynamic society than ours. As mentioned earlier, scientific progress might, in the long run, bring such an extension of the individual life span that it would seem like a kind of immortality, at least in a relative sense, since accidental death could not in any case be avoided.

Drake deduces that a species that has reached this stage will necessarily give an enormous value to life and therefore will check its own aggressiveness so as not to put it in danger. Further, it will be interested in defusing the aggressiveness of the species it will encounter, teaching them the techniques of life-span increase. One of the first messages we might receive from a species of this kind, therefore, would be a sort of recipe for the elixir of life.

No doubt it is a clever hypothesis, but it is not free of weaknesses. First, a species of "immortals" might give an extremely high value to their own lives and will do everything not to put life at stake, but this does not guarantee that that species would attribute an equally high value to the lives of others. From a purely logical point of view, would it not be safer to exploit their own technological superiority to eliminate all other intelligent species before they can become a danger, or at least to shut themselves in their ivory tower? Then there are those who say that an immortal would be so fed up with an overlong life as to put it voluntarily at risk in a violent action, and hunting primitives might turn out to be their favorite sport. This is not to say that Drake's hypothesis has no value, but just to show that, with our present knowledge, any statement on this subject can be contradicted by an opposite statement, without it being possible to choose between the two through purely scientific reasoning.

Even the statement that a civilization so ancient, and therefore so technologically advanced, must be also morally at a higher level than ourselves may not have a very good basis. Humankind did not change much from this point of view since historical sources existed, and technological progress and moral progress can go at different paces. Speaking of extraterrestrial intelligence, the very concept of moral progress may be too anthropomorphic. Also, when speaking of conscious *and* intelligent beings, we must assume that they are individuals, each one with his own free will, whatever such a concept may mean in the present context. Classifying a civilization or an intelligent species as more or less morally advanced (therefore classifying it as good or evil) is another senseless generalization, like so many others that have brought tragic consequences throughout human history. If, for instance, a hypothetical extraterrestrial civilization had to decide whether to contact us or not, would it judge us by looking at the moral codes of our most common religions and philosophies, or at the behavior of the countless humans who infringe such rules? This is not only true at the level of individual behavior, but also of collective behavior, as testified to by the many crimes committed—even very recently—by states or other political entities.

Surely these considerations go beyond the scientific arena and trespass into philosophy or religion. The Bible gives an answer to this problem: if we accept that Original Sin is universal, involving all nature throughout the cosmos, then all intelligent species have in themselves a mixture of good and evil and it makes no sense to define a species as good or evil as such. The balance between the tendencies toward good and evil present in every individual is a dynamic equilibrium, varying in space and time, as the history of our own species teaches us. Since human nature is extremely complex, rules will have to be established to regulate relationships between species; relationships with intelligent beings of extraterrestrial origin must, sooner or later, become a judicial matter.

LEGAL RULES FOR RELATIONSHIPS BETWEEN INTELLIGENT SPECIES

Metalaw is a term coined by Andrew G. Haley in 1956 to define a system of laws that could apply to all relationships between intelligent aliens. But no such metalaw is contemplated in the legislation of any country.

At present no general definition of intelligent being is contemplated by law. But if this is the case, what could possibly be the legal status of a hypothetical intelligent alien who lands on our planet? Clearly it depends on the country he visits, since in some countries even humans enjoy quite limited rights. In the past humans were treated very differently depending on their race, citizenship, religion, sex, status (whether free or slave), etc. One of the conquests of modern civilization is precisely that of recognizing equal rights to all humans. The basic point is the concept of "person," but it is debatable whether an extraterrestrial would be included in this definition. It has been pointed out that in the United States an alien would at least be legally recognized as an animal: "a living being, not human, endowed with the power of voluntary motion." Whether it would be considered a "wild animal" or something like a household pet is a matter for lawyerly discussion. At any rate, this would at least grant the alien animal rights to "adequate housing, ample food and water, reasonable handling, decent sanitation, sufficient ventilation, shelter from extremes of weather and temperature"!

In his paper *Metalaw and Interstellar Relations*,[30] Robert Freitas reports two possible ways in which an alien could have its status of "person" legally recognized. The first is the possibility of defining "moral agents" or "moral persons," that is, beings who can take moral stands or make moral-ethical judgments, regardless of whether they coincide with ours or not. The other is that of defining self-aware beings—that is, beings who feel themselves separate from their surroundings. Moral persons or self-aware beings can thus be considered "persons." Clearly both definitions leave much to be desired; they would be quite difficult to apply in practice and may be much too anthropomorphic. The principle of reciprocity should at any rate apply, so the alien should be considered in the same way that a terrestrial human is considered by aliens.

Once the alien is considered a person, a basic rule is that he should be treated as he desires to be treated. The basic rule that you should treat other people as you would like to be dealt with by them is much too anthropomorphic and in the context of conducting ourselves with aliens could lead to misunderstanding and even physical damage. Fortunately, relationships with aliens are likely to occur not on a personal level, but as a general relationship between communities. A basic political rule is the non-interference principle, well known to *Star Trek* fans as the Prime Directive. It sounds like this:

> As the right of each sentient species to live in accordance with its normal cultural evolution is considered sacred, no Star Fleet personnel may interfere with the healthy development of alien life and culture. Such interference includes the introduction of superior knowledge, strength, or technology to a world whose society is incapable of handling such advantages wisely. Star Fleet personnel may not violate this Prime Directive, even to save their lives and/or their ship, unless they are acting to right an earlier violation or an accidental contamination of said culture. This directive takes precedence over any and all other considerations, and carries with it the highest moral obligation.

The full text of the Directive has been quoted because, if strictly adhered to, it would forbid any contact between civilizations except in the unlikely case of civilizations at exactly the same cultural level. No two

[30] http://www.rfreitas.com/Astro/MetalawInterstellarRelations.htm.

civilization would probably ever be considered such, so the Prime Directive actually forbids any communication. Who could decide whether a civilization must remain in isolation? While a weak form of the non-interference principle might be a wise basis for relationships, its consequences should not be carried to the point of preventing all contacts.

The basic principles of metalaw were synthesized by Ernst Fasan in 1970 in 11 rules that constitute the basis of an interstellar policy that should bind all intelligent beings:

1. No partner of metalaw may demand an impossibility.
2. No rules of metalaw must be complied with when compliance would result in the practical suicide of the obliged race.
3. All intelligent races of the Universe have in principle equal rights and values.
4. Every partner of metalaw has the right to self-determination.
5. Any act that causes harm to another race must be avoided.
6. Every race is entitle to its own living space.
7. Every race has the right to defend itself against any harmful act performed by another race.
8. The principle of preserving one race has priority over the development of another race.
9. In case of damage, the damager must restore the integrity of the damaged party.
10. Metalegal agreements and treaties must be kept.
11. To help the other race by one's own activity is not a legal but a basic ethical principle.

Such rules are without doubt a good starting point on which to build the laws and ethics of relationships between species, but they have been elaborated by one of the sides only—and it could not be otherwise, since it is not even certain that the other sides exist. Moreover, if it is true that the other species with which we could come in contact are much more ancient than ourselves, it is likely that they already faced this problem and established rules for relationships among species.

INTERSTELLAR PREDATORS

Any law implies that the parties who decide to comply with it reciprocally recognize their rights, as explicitly stated by point 3 of the metalaw. However, there is no guarantee that all intelligent species acknowledge the dignity and respect that we assume all intelligent and conscious beings deserve from each other. At the lowest level, any living being may simply be considered a source of food and energy by any other living being based on a compatible biochemistry. On our planet, after all, the relationships among species are usually dictated by the prey–predator antinomy.

It has been suggested both that an intelligent species is unlikely to be autotrophic and that the development of the human brain was made possible by a diet that is at least partly carnivorous. The complex interactions among the members of a band of hunters are also considered to be an important factor in the development of intelligence and even of consciousness. The food chain on our planet is quite complex, but we can assert that at the lowest level we have life-forms (plants) able to synthesize the organic substances they need through photosynthesis using solar energy. Plants are used as food by other organisms, which can in turn be eaten by others and so on. It is true that the efficiency with which solar energy is utilized decreases along the food chain, but feeding on organic substances already prepared by other organisms, or by other animals, allows a compact and readily usable energy source to be exploited. As an added bonus, all substances essential for complex life, even if needed in very small quantities, are easily obtained.

Humans are high in this food chain, and it is reasonable to expect that other intelligent beings are also in a similar position. In the case of humans, the diet and food habits also depend on cultural traditions and personal taste, and the same may apply to other intelligent beings. It is then reasonable to wonder whether the biosphere of a planet may become an interesting source of food and energy for beings living on another planet. Charles Cockell and Marco Lee tried to answer this question in a paper published on the *Journal of the British Interplanetary Society*.[31] Their conclusion is that the energy content of the biosphere of a planet may, at least in certain conditions, be large enough to justify the interstellar journey, so it is possible to imagine that there are beings who get food and energy in this way: they may be defined as interstellar predators. We can assume that they are intelligent beings, since it is reasonable that intelligence is needed to undertake interstellar journeys.

A planetary biosphere must be quite rich in energy to justify such long journeys, but difficulties are actually even greater. Biochemistry on Earth is based on sugars with chirality D (dextrorotatory) and amino acids of L type (levorotatory)—sugars and amino acids of different types have a reduced nutritional value, or rather their value depends on whether

---

[31] C. Cockell and M. Lee, "Interstellar Predation," *JBIS*, vol. 55, pp. 8–20, 2002.

mechanisms are available to decompose substances that have the "wrong" chirality. Mice, for instance, can use levorotatory fructose. Moreover, fats have no chirality whatsoever and can be used without this limitation. If the chirality of terrestrial biochemistry is due to cosmic factors, other biospheres based on sugars and amino acids will have the same characteristics and interstellar predators will be more likely. If chirality is random or due to reasons linked with our planet, the nutritional value of an alien biosphere could be much lower. One can also imagine a truly omnivorous predator, able to decompose and metabolize substances of all types.

But aside from these details, is the hypothesis of interstellar predator species reasonable? Many doubts can be raised on this issue. It has been noted that human species passed through a neolithic revolution during its cultural evolution, and learned to produce its own food by taming plants and animals, and abandoning an economy previously based on hunting and gathering. Such a neolithic revolution is probably a compulsory passage in the cultural evolution of all intelligent species. Today we are likely to be close to a further revolution, in which humans will learn to produce their food directly from nonbiological matter instead of using tamed animal and vegetal species. It is reasonable that, before being able to attempt interstellar journeys—a step needed to become interstellar predators—an intelligent species becomes technologically autotrophic and no longer needs to feed on other species. This reasoning allows us to think that the hypothesis of interstellar predators is just a good idea, and a much frequented one, for science-fiction novels and movies, in the style of *Independence Day*.

ENCYCLOPEDIA GALACTICA AND GALACTIC INTERNET

A community of different intelligent species in contact with each other has often been predicted and termed the *Galactic Club*. Establishing contact with an extraterrestrial species would therefore give humankind the possibility of entering this community, accepting its rules, and enjoying some of the benefits reserved to its "members." This possibility must be put in a correct perspective, taking into account the distances we must deal with when speaking of interstellar relationships. Given these distances, what might the activities of this Galactic Club be?

So far only contacts at a distance—possibly using radio waves, at least at our present technological level—have been considered. However, regardless of the means used (radio, laser, neutrinos, etc.), the limitation of the speed of light for the transmission of information holds, so communication with extraterrestrial intelligence could hardly take the form of a dialogue. Only if there were many intelligent species in our galaxy could one of them be at a distance of, say, 30 light-years or less, so that less than 60 years could pass between transmission of a message and reception of the answer. Although if that were not the case, it will still be possible to speak of a dialogue between species, if extremely slow, but surely not of a dialogue between individuals.

This consideration may be mitigated by a significant increase in the duration of human life, since beings with a much longer life span could have a very different perception of time and be interested in a dialogue in which the intervals between a question and an answer are very long. However, the most likely distances between intelligent species are of thousands of light-years, or even more, and it's hard to imagine a species whose individual life is long enough for them to be interested in a dialogue that had such excruciatingly long pauses.

The typical life spans of the various species depend on the physico-chemical processes on which they are based. If we think the biochemistry of other species is similar to ours, then life spans cannot be much different, either, and will be determined by the time required by the chemical reactions on which the workings of our body are based. Other intelligent beings might be faster or slower than ourselves, but not by orders of magnitude. If, on the contrary, they are based on different principles, much faster (like the processes occurring in computers) or slower (like chemical reactions occurring at low temperature), their view of time could be radically different from ours. Consider, for instance, intelligent beings based on reactions that take years to complete. Their life could last billions of years, even if subjectively they would find it no longer than ours. If they are able to achieve interstellar travel, for them the speed of light will be a negligible limitation and a dialogue by radio with a planet of another star would only introduce what, to them, would be a short delay of the answer. To us they will look like inanimate beings, unchangeable as (and even more than) rocks, and they would probably also not be aware of our presence without using instruments, in the same way that we cannot follow fast phenomena without suitable instrumentation.

So owing to the characteristic times of our biochemistry, the dialogue we may have with other species will probably be reduced to something very similar to a series of monologues. The most interesting perspective from this point of view is that an intelligent species broadcast the *summa* of its knowledge, putting it at the disposal of other species. In case many species are in contact with each other (the Galactic Club), it might be a synthesis of knowledge accumulated by the whole community of intelligent beings. Using a term introduced by Isaac Asimov in a very different context, a transmission of this kind is usually referred to as an *Encyclopedia Galactica*.

Such a message obviously has an enormous interest, even if there are legitimate doubts that it could be decoded and understood. Even if the statement by Minsky that a common ground exists for the various intelligent species to understand each other, the difference of cultural level could make a transmission of this kind as useful to us as the finding of a modern physics textbook by a Neanderthal man. More than an *Encyclopedia Galactica*, we could perhaps use their primary school textbooks! In any case it would be interesting to have an idea of how much information an encyclopedia of this kind could contain and how much energy would be needed to broadcast it at interstellar distances in such a way that it arrives on Earth with enough power to be detected by our receivers.

Recently a new idea was added to the concept of an *Encyclopedia Galactica*: the *Galactic Internet*. Starting from the hypothesis that travel at interstellar distances will remain extremely difficult and expensive forever and that communications by radio, if based on transmissions at very large distances with coverage of large zones, require very high power, Timothy Ferris proposed the hypothesis that civilizations that exist now or existed in the past in our galaxy built a net based on a great number of small probes in the various star systems that act as relays, exchanging information among themselves by radio. Every civilization will introduce information about its science, culture, and history in the net, information that would spread at the speed of light and remain located in the various nodes, at the disposal of whoever wants to log on.

Every new user—intending by such a term a new civilization that logs onto the net—would therefore build a new node in its system (if there is none already built) and eventually launch probes in the nearby systems to extend the net. The new nodes would receive all information from those already connected and introduce new data, that would spread around and remain at everyone's disposal. Any user of the system could connect to the nearest probe and unload all the information he requires, as from any Internet site.

In this way access time would be drastically reduced, since the user would only see the probe located in his planetary system, a few light-hours from him, at the cost, however, of an enormous multiplication of the memory required; every probe would contain all the information existing in the net. It is, as already said, a technologically more advanced variation of the concept of the *Encyclopedia Galactica* that allows a certain interactivity, since the time of access is reduced from the years or centuries needed for interstellar communication to the few hours needed to contact the nearest probe. For something of this kind to be possible, it is essential to build information systems that are able to store quantities of information greater by many orders of magnitude than those currently used by our civilization, but this is not a severe limitation of this concept, judging from the pace at which the power of our computer systems is growing.

A net of this type would permit the retention of all the information on extinct species and therefore to overcome the problems due to a limited duration L in Drake's equation. The greatest problem is the maintenance of the sites, both as concerns the duration of the probes and their computer systems, and where management of the information is concerned. If the intelligent species that enter the net are many, and if updating is frequent, then it is not practical that all information should be stored forever—as that would require an indefinite extension of the memory needed. The probes, therefore, should be endowed with a high level of artificial intelligence, to the point that they would be similar to the Von Neumann probes described earlier.

THE ZOO HYPOTHESIS

The fact that no contact with extraterrestrial intelligent species has yet occurred, in spite of decades of attempts and a certain visibility of our civilization for more than half a century, caused the formulation of a number of hypotheses asserting that any intelligent species that was aware of our presence avoided any contact on purpose. These hypotheses also try to make the diffusion of intelligent species in nearby zones of our galaxy compatible with the possibility of travel and communication at interstellar distances, sometimes in very fanciful ways and always without any possibility of experimental validation. Since they postulate the impossibility of experimental verification, and are therefore not falsifiable, they cannot be considered scientific theories.

The term *zoo hypothesis*, used here to designate such hypotheses collectively, comes from the idea that the condition of humankind on Earth is a sort of anthropological experiment performed by alien scientists, who set the first humans on this planet (or waited for an intelligent species to evolve autonomously on Earth) and are now observing our behavior, like zoologists would observe the behavior of the animals in a cage at the zoo. Obviously, the scientists who are following the experiment refrain from interfering, since any interference could affect its results.

Endless variations of this hypothesis exist, the first one probably formulated by Tsiolkovsky, who thought that many intelligent species exist in our cosmic neighborhood but avoid any interference with our development, fearing that contact with an evolved species could damage us, as happened on Earth when less advanced civilizations came in contact with technologically more advanced ones, or may at least cause terrestrial civilization to lose the characteristics that differentiate it from the others. If this happened, the earthlings could no longer bring an original contribution to the galactic civilization when, after developing the technologies needed for interstellar travel, they enter this larger community. Note the similarity of this hypothesis with the already mentioned "Prime Directive."

The problem with these hypotheses is the impossibility of conceiving an experiment that distinguishes a scenario where extraterrestrial intelligence does not exist from one in which it exists but deliberately avoids contact, whether as an act of philanthropy, of political foresight, or as a scientific experiment. In the absence of experimental verification these are just nonverifiable conjectures, weakened by a good deal of anthropomorphism, which do not merit further discussion.

INTERLUDE:

THE SEARCH FOR
EXTRATERRESTRIAL
STUPIDITY

EVERY intelligent being interacts with other intelligent beings through his actions; these interactions are bound to cause advantages or damage (in terms of material or nonmaterial gains and losses) both for the agent and for its counterparts. An interesting method for evaluating human behavior, based on the gains and losses caused by these interactions, was proposed by Carlo Cipolla in his essay on the basic laws of human stupidity.[1] The aim of the present interlude is to verify whether conclusions of the same type might be applied not only to humans of our planet but to all intelligent beings. The basic instrument is a

---

[1] Carlo M. Cipolla, "Le Leggi Fondamentali della Stupidità Umana," in *Allegro Ma non Troppo*, Il Mulino, Bologna, 1988.

plot of the type shown in Figure 1. On the horizontal axis are gains (losses are on the negative part of the axis) the actor causes to himself, while on the vertical axis the gains and losses he causes to the individual with whom he interacts. If the interactions are many, as happens when we want to study the behavior of a given actor with respect to many others, the average value of the gains and losses in the various interactions must be reported.

If the representative point corresponding to the gains and losses caused by the actions of a given person lies in the first quadrant (the interaction gives advantage to both the actor and the others) that person is defined as *intelligent*. If it lies in the second quadrant, that is, the action gives advantages to the actor but advantages to the others, the actor is an *unwary* person. If it lies in the fourth quadrant, that is, the actor obtains advantages that cause disadvantages to others, we face what Cipolla calls a *bandit*. The simplest case is that of the thief, whose action lies on the bisector of the fourth quadrant (line OD, the gain he gets is equal to the loss of others) only if the theft does not cause collateral damages.

Finally, those whose actions lie in the third quadrant, acting in such a way to be a nuisance to themselves and others, are *stupid*.

Intelligent, unwary, bandit, and *stupid* are then the four possible types of behavior of human beings toward their fellow humans.

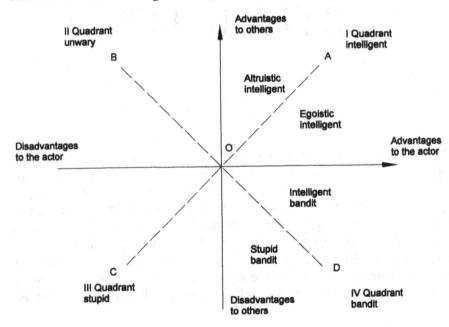

FIGURE 1 *Plot showing the gains and losses an actor causes to himself and to others.*

As a first conclusion, if a being acts completely at random, statistically the gains and losses should compensate and the characteristic point should be the origin O. However, all living beings are the result of evolution, which cannot avoid favoring a behavior that gives advantages to the actor without causing disadvantages to the other beings of the same species. Moreover, a behavior that gives advantages to the other beings of the same species should also be of advantage to the actor and be even more favored by evolution. As a consequence, a nonintelligent being acting in a way dictated only by his instinct should have a behavior that, in this context, has been defined as *intelligent*, perhaps as an *intelligent egoistic being* (the term *egoist* must here be intended in its "good" sense, meaning that it seeks its own gain without bothering to benefit others, but without damaging them).

It follows that only an intelligent being (in the sense of a being free to behave in a way not rigidly determined by instinct) can be *stupid*. This conclusion can be generalized, obtaining the following Universal Stupidity Principle (USP):

Only a being belonging to an intelligent species can be stupid.

This principle can be stated in a weak form, which interprets the previous statement as a necessary condition:

A stupid individual necessarily belongs to an intelligent species.

A strong form, considering the statement both necessary and sufficient, is:

To include stupid individuals is a characteristic of all intelligent species, and only of intelligent species.

The strong formulation seems to be more consistent and therefore will be adopted here.

No doubt this approach is too anthropomorphic, but not much more anthropomorphic than most other considerations usually made when dealing with extraterrestrial intelligence, or with extraterrestrial life in general. Moreover, it is perhaps possible that an intelligent species free from stupidity exists, but it must be an intelligence so different from ours as to be unimaginable for us, and we would probably not recognize it as such.

If the USP holds for all intelligent species, the four laws of human stupidity formulated by Cipolla become universal.

The Laws of Universal Stupidity

1. Every one of us always and invariably underestimates the number of stupid people hanging around.
2. The probability that a given person is stupid is independent of any other characteristic of that person.
3. A person is stupid if he/she causes losses to other persons or group of persons, without obtaining any gain, or even causing a loss to him/herself.
4. Nonstupid persons always underestimate the potential danger of stupid people. In particular, they constantly forget that to deal with or associate with stupid individuals in any place or time will inevitably prove to be a very costly mistake.

What are the consequences of this principle and the associated laws?

First, we must note that, in general, aliens described by science fiction do not comply with these laws. They mostly behave in ways that, on the basis of Figure 1, would qualify them as *intelligent* or *bandits*. As an example, the aliens of *Independence Day* and of dozens of other books and movies, starting with *The War of the Worlds* by H.G. Wells, are surely *bandits*. On the plot they lie on the bisector OD of the fourth quadrant, since they try to steal our whole planet from us, taking away our existence to obtain their survival. In the "terrestrial chauvinism" of Hollywood scriptwriters they may look *stupid*, since in the end they do not get the advantages they look for, and succumb. Perhaps the message of these novels and movies is that any aliens who fight against the humans of planet Earth are *stupid*, if nothing else because they are doomed by the need of a happy ending; but it's really more consistent to define them as *unlucky bandits* than *stupid*.

Many other aliens are benevolent creatures, who come to Earth for our own good and obtain very little profit from their actions. Not only are they *intelligent*, they are *altruistically intelligent*, very close to the vertical axis in Figure 1. In some cases their altruism could make them look *unwary*, since they experience trouble and loss to save humankind without obtaining anything for themselves, but those who behave in this way should be classified more among the *saints* than the *unwary*.

Stupid aliens are seldom seen in novels and movies, and slightly more often in space operas than in science fiction, except when the author wants to create comic characters. Moreover, the few stupid aliens are usually very anthropomorphic.

Extraterrestrial intelligent beings, as SETI scientists usually imagine them, violate the laws of universal stupidity to an even greater extent. The very statement that the alien civilizations we could come in contact with will be older and wiser than ours and will share their technology and wisdom with us to defuse the potential dangers coming from a young civilization, assumes that these aliens are *intelligent,* that they would lie in the first quadrant of Figure 1, probably close to the bisector OA. This statement clearly violates the second law, which implies that the probability that a person is *stupid* is independent of the evolutionary stage of the intelligent species to which he belongs. A Neanderthal man had the same chance of being stupid as a Cro-Magnon man, that is, a man of our own species.[2] Regardless of how evolved an intelligent species is, we must expect potentially stupid behavior from its members, which is liable to produce losses for us without producing any gain for them (third law). In all contacts we must therefore exert the greatest caution, which will nevertheless not be sufficient (fourth law). The fact that—applying the fourth law in the opposite direction—the intelligent ones among the aliens will necessarily have to deal with some stupid Earthlings will not suffice to solve the problem, since the stupid Earthling will produce losses for them without obtaining any gain to compensate for the losses due to the aliens.

As Cipolla correctly stated in his essay, stupidity is far from being a zero–sum process (the gains on one side cannot compensate for the losses on the other side) but causes a global loss, shared in different ways among the actors of the interaction, in the same way that an intelligent behavior generates gains for all parties.

Individual stupidity plays a role in the relationships among different intelligent species similar to that played in relationships among individuals of the same species—a role that is dangerous for all species. Each intelligent species in the Universe must exercise the utmost care in choosing the individuals to whom it entrusts the task of conducting the interaction, since a high presence of stupid individuals would cause losses on a cosmic scale.

From this viewpoint, the first and fourth laws can have very severe consequences and lead to the most pessimistic conclusions. Many fear that human stupidity will ultimately lead to the extinction of our species, and this conclusion can easily be extended from our planet to the whole Universe.

The situation bears some similitude to the consequences of the second

---

[2] As already stated, many anthropologists held that a Cro-Magnon man could not be distinguished from a modern man, if not for cultural differences. It must be explicitly noted that the Laws of Universal Stupidity cannot be applied to hominids or to *Homo habilis, Homo ergaster,* etc., who were not fully self-conscious—in other words, they were not intelligent enough to be *stupid.*

law of thermodynamics: the tendency of entropy to always increase can lead to the eventual death of the Universe. However, one hope can perhaps be cultivated. Recently the hypothesis has been advanced that some still unknown physical law will keep the increase of entropy in check, thus avoiding the final death of the Universe. In the same way, unknown physical laws could exist posing a limit to the damage caused by stupidity, and ultimately allowing intelligent species to survive their own stupidity.

5

THE POSSIBILITY OF
CONTACT

THE FERMI PARADOX

SOME witnesses reported that in 1950 Enrico Fermi was discussing the possible existence of extraterrestrial intelligent beings with Edward Teller, Herbert York, and Emi Konopinski in Los Alamos. His opinion having been asked on the matter, he skeptically replied that he did not see any around. It sounds like just a joke, but actually it is a much more serious objection than may be apparent at first sight. Fermi was of the opinion that very long distance space flight will be possible (actually, he thought superluminal space travel was also likely), at least in the distant future, and that this possibility holds for any sufficiently technologically advanced civilization.[1] Since nobody has ever supplied convincing

---

[1] A detailed description of the circumstances in which Fermi formulated the paradox is reported in S. Webb, *If the Universe is Teeming with Aliens. . . Where is Everybody?* Springer, 2002. The book is fully dedicated to the Fermi paradox and discusses fifty possible solutions.

evidence of the presence of extraterrestrials on our planet, we must conclude that they do not exist.

The existence of intelligent species in our galaxy, therefore, would be an alternative to the *space imperative* and above all to the *conscious life expansion principle* or, at least, would imply the existence of physical limitations to the expansion of an intelligent species in the Universe. It has already been said that if other intelligent species exist in our galaxy, most of them must be much older, and therefore technologically more advanced, than we are. If the *conscious life expansion principle* also holds for them, then by now they should have colonized, or at least visited, the whole galaxy and therefore we should have already met them a long time ago.

The first one to encounter this problem was Tsiolkovsky, who believed both in the existence of a plurality of intelligent species and in the possibility of interstellar travel. However, as his philosophical works only recently became known in the West, the matter is generally known as the Fermi paradox or the Fermi question. It may be generalized to automatic probes, since some of those who deny the possibility of interstellar journeys hold that it will be possible to colonize the whole galaxy (or even the whole Universe) using self-replicating robotic probes, the so-called von Neumann probes.

Still, the question posed by Fermi cannot be easily dismissed. The basic solutions are essentially of three types:

- Many people do not accept the statement that extraterrestrials are not here, from supporters of the existence and extraterrestrial origin of UFOs, to those who think that we have not yet seriously looked for traces of extraterrestrial visits to our planet.
- There are those who suggest that either extraterrestrial intelligences do not exist or, if they exist, intelligence is such a recent phenomenon that they could not yet have reached our planet.
- The last solution is, of course, that interstellar journeys are impossible, both for intelligent beings and for their automatic probes.

A surprising number of supporters of astronautics deny the existence of extraterrestrial intelligence and many supporters of SETI deny the possibility of interstellar travel. Actually tens of alternative hypotheses may be formulated, like the zoo hypothesis mentioned earlier. Tsiolkovsky for instance, believed that intelligent species, much more advanced than we are, avoid letting us know of their existence because premature contact could damage us and at the same time prevent the galactic community from receiving the original contribution to global civilization that we could give, if left free to develop autonomously.

Yet the Fermi question can also be answered without resorting to fanciful hypotheses and without dismissing one of the two alternatives. Ivan Almár[2] suggested that the Fermi paradox has some points in common with the Olbers paradox, stating that if we are in an infinite and static Universe, the sky should be bright at night, instead of being dark. It has been suggested that the night sky is dark because of interstellar matter intercepting the light from distant galaxies. This is not really a good answer, since interstellar matter absorbing so much energy would radiate itself and contribute to the overall luminosity of the sky. Only modern astrophysics realized that the two causes are the red shift of the light from distant galaxies and, above all, the limited value of the speed of light. Owing to the latter, the light from very distant galaxies simply has not reached us yet and the radius of the visible Universe cannot be larger, in light-years, than the age of the Universe in years. But the Olbers paradox will not really be solved until a better knowledge of the Universe is gained. Almár also suggests that we may not yet have all the knowledge required to solve the Fermi paradox. Some factors slowing down the expansion of intelligent species might exist, as well as other explanations of which we are not yet aware.

It is in fact possible to imagine scenarios where intelligent species are sufficiently far from each other in time and in space and the time needed for their expansion in the galaxy is long enough to explain the lack of contacts. As already mentioned, with the purpose of solving the Fermi paradox the hypothesis has also been advanced that our species is really one of the oldest in our galaxy, perhaps owing to gamma-ray bursters that prevented older species from developing intelligence.

What really is impossible to reconcile are the extreme positions of those who think that intelligent species are extremely common and those who believe that it will be possible to colonize the whole galaxy in a relatively short time.

As noted by Stephen Webb in his book devoted to the Fermi paradox, each scientist who has dealt with it has his favorite explanation. Webb's is that, for a number of reasons, intelligent life is so uncommon that we are probably unique in our galaxy. The author of the present book thinks that the expansion in a galaxy is a slow process, much slower than can be computed by simply dividing the distances by the travel time; a species must not only cross a galaxy but settle it, with billions of potential planets to explore, to terraform them when they are

---

[2] I. Almár, "Analogies Between Olbers' Paradox and the Fermi Paradox," *Acta Astronautica*, vol. 26, no. 3/4, 1992, pp. 253–256.

too hostile, and to colonize them to the point of being ready to mount the next expansion expeditions. Given a limited number of intelligent species, the chances that one of them has already reached our system may well be quite low.

As a last consideration on the Fermi paradox, it is unwise to take it too seriously, considering it as proof that we are alone and then giving up SETI, or concluding that interstellar travel is impossible and giving up all hope of becoming an interstellar civilization. There are so many unknown factors and so much wild guessing in all that has been said about the Fermi paradox that any solution we can devise is nothing other than conjecture.

UNIDENTIFIED FLYING OBJECTS

Many people believe that extraterrestrials are already among us and that the frequent sightings of unidentified flying objects (UFOs) are evidence that their space vehicles periodically visit the Solar System and Earth's atmosphere. Some polls show that in certain countries, such as the United States, most people find this very likely and a substantial minority deeply believe in the extraterrestrial origin of UFOs. Besides, such a belief feeds a rich market of books, lectures, objects, etc., the world over and is therefore also reinforced by strong economic motivations.

The hypothesis that somebody else has developed the technology to perform interstellar journeys cannot be lightly dismissed and it cannot be excluded that the first contact with an extraterrestrial intelligent species will occur on Earth, with the arrival of an alien spaceship. After all, this is exactly what the Fermi paradox would require. On the other hand, the author and most scientists believe that there is no convincing evidence of the presence of extraterrestrials in the Solar System in general and on Earth in particular.

Most of the UFO sightings have been explained with a variety of natural phenomena or, in some cases, with deliberate hoaxes, even if a certain number of cases do exist for which no convincing explanation has yet been found. For these cases the above-mentioned criterion, according to which exceptional discoveries require exceptional evidence, must be applied, and at present no one has been able to supply really convincing proof of the extraterrestrial origin of these objects.

Very often those who believe that extraterrestrials visit our planet hold that the civil and, above all, military authorities of all countries join in orchestrating a plot aimed at preventing the public from knowing such a disturbing truth. As evidence not only of the presence of extraterrestrials on our planet but also of the fact that both American and Russian experts studied their vehicles, the rapid progress in the field of aeronautics and astronautics of the 1960s, culminating with the landing on the Moon in 1969, is sometimes mentioned.

This so-called evidence does not hold up to even a summary analysis: the developments of the 1950s and 1960s are a direct consequence of the progress achieved in the previous decade under the pressure of the war, and there was nothing really revolutionary in the *Saturn V*, the rocket that took the Americans to the Moon. From the conceptual point of view, it is nothing other than an enormously enlarged and upgraded *V-2*. The materials used for conquering the Moon are, to a large extent, the same ones that were used to build the aircraft of World War II and it is possible to follow the logical lines of the technology's development and see that there are no sudden jumps or external contributions of any significance. If the parties engaged in the arms race that characterized the Cold War really could have had access to an incomparably more advanced technology, they could not have avoided using it, with consequences impossible to keep secret. Even the better known cases, like the Roswell accident, did not stand up to deeper investigations, and the certainties that many boast about these extraterrestrial visits have the same scientific foundations as other pseudo-sciences, such as astrology.

Other events connected with the sighting of UFOs are the so-called abductions of humans by extraterrestrials. It is quite a complex phenomenon, which in many cases cannot be simply dismissed as mythomaniacs telling tall stories, or as interesting inventions. There is no doubt that these aspects are present and that, in this case too, the market for these stories in books and magazines, if nothing else, well explains frauds and inventions. But the fact is that many of the people who are convinced that they were abducted, brought on board alien space vehicles, and used as guinea pigs for more or less scientific experiments, believe it in good faith. There is a vast literature on the matter, by abductees and by researchers, particularly psychologists and psychiatrists, who studied the persons involved in these events. An interesting account of this phenomenon can be found in the book by Clifford Pickover, *The Science of Aliens*.[3]

None of the abductees, however, has ever been able to bring back from his experience even the slightest evidence, not to speak of the exceptional evidence that such an exceptional event would require. Besides, extremely dubious circumstances abound: the description of the aliens changed in

---

[3] C. Pickover, *The Science of Aliens*, Basic Books, New York, 2000.

time, from the 1940s to the present, to settle now on a stereotyped image of humanoids with a big head and two very large eyes without iris or pupil (Figure 5.1). Yet the descriptions differ enough from one another to exclude the beings from belonging to the same species. How is it possible that the various intelligent species that come in contact with us are similar to us, are so abundant, and, depending on the time of their arrival, are also similar among themselves? It has been mentioned that the aliens described recently are markedly similar to human fetuses. On this matter Michael Grosso, quoted in Pickover's book, points out that they also have a striking similarity with the images, repeatedly shown by television, of famine stricken children, with their large eyes and a big head on a skeletal body.

Another dubious point is the fact that those who said they had been abducted often reported that they had witnessed the results of interbreeding experiments between aliens and humans, or that they had been subjected to experiments of this type, such as the removal of ova from their body. According to these stories it would seem that the main reason for which these aliens come to Earth would be essentially to get human genetic material and, often, human ova to fertilize with alien sperm to create a hybrid species! But the cross-fertilization between different species, even if phylogenetically very close and evolved in similar environments, is impossible. For surely aliens belong to a species very different from ours, and the hypothesis of any cross-fertilization is, one must assume, unthinkable.

These absurdities, together with the absolute lack of material evidence, make us think that the experience of the abduction by aliens happens entirely within the minds of the abductees. Many of them describe

FIGURE 5.1 Sketches of aliens reconstructed on the basis of the descriptions supplied by abductees. (Sketches inspired by the faces of extraterrestrials reported in the book by C. Pickover.)

symptoms that have often been interpreted as typical of epilepsy, and many instances of abduction have been explained with epileptic forms, particularly epilepsy of the temporal lobe. Not all agree with the generalization of this explanation, because similar forms could not be diagnosed in all the abductees, but experiments in which experiences of the same type (the impression of hands that grab one by the shoulders, etc.) have been induced by the application of electric fields to the brain, are considered important confirmation of the fact that we are facing experiences occurring completely inside the mind of the subjects involved. On the other hand, today we see a flourishing of fantastic stories about unbelievable facts and "urban myths" told as true facts, often by people who make such statements in good faith; the press and television give an exaggerated relevance to these facts, which is proportionate only to the audience they usually attract.

Finally, it must be noted that abductions by aliens are experiences comparable to those described in the past, when some people stated that they had been abducted by demonic (or angelic) creatures or had relationships of various natures with them. Each culture and each historical period had well-defined stereotypes of the creatures involved in these events, and today it seems that aliens with large eyes in a big head on a very thin body are having their moment. It is worthwhile to remember the extreme caution of the Christian Churches in dealing with visions and miracles of all types—a caution many journalists, writers, and television anchors would do better to imitate.

Nevertheless, those who think that the Fermi question must not necessarily be answered by denying any of the terms of the problem, must consider with open minds the possibility that the contact between our species and extraterrestrial intelligences might occur on our planet, with the arrival of an alien spaceship. If there is no evidence that this happened in the past, and therefore the description of contacts must be considered as a product of imagination, the future might present surprises in this area and every sighting must be investigated without prejudice.

HISTORICAL TRACES OF PAST ENCOUNTERS

One of the possible solutions to the Fermi question is that extraterrestrials visited our planet in the past. This possibility cannot be dismissed. If, as has been said several times, other intelligent species, much more ancient than humankind, live in our galactic neighborhood, close to the Solar System, it seems unlikely that one of them should make an exploration journey exactly now, when we are beginning to explore space. This lack of contemporariness may be of hundreds of thousands or, more likely, millions of years, and therefore the probability of finding meaningful traces of a short visit of a small group of extraterrestrials to our planet is vanishingly small.

Yet there are many who think that history is rich in traces of extraterrestrial interferences and bring a large body of "evidence" to support their claim. The majority of such proofs, which could at best be considered weak clues, originate from written texts and therefore date back to a maximum of a few thousand years. Very often they are based on the interpretations of obscure texts, generally religious texts, and even many passages from the Bible have been used in this way. Apart from the many interested frauds, there are often clever, at times fascinating, interpretations that stir remarkable interest in the public. In some cases the interpretations are mixed with scientific problems that are still open, in a context where the former yield a key to solve the latter, which in turn supply validation of the first, in a cycle in which certainties are drawn by putting together facts and very doubtful clues.

An interesting example is the theory according to which extraterrestrials colonized our planet about 200,000 years ago, performing genetic engineering operations on the hominids. By inserting part of their genetic material in our ancestors, they would create intelligence on Earth, to rule like gods[4] on the creatures so generated. In this way the scientific problem of the origin of intelligence is solved, and an interpretation of many legends involving various divinities is supplied. Even the biblical sentence "God created man in his own image" is elegantly explained!

Other evidence of the arrival of extraterrestrials is obtained from sculptures and drawings in which some objects are identified (from space suits to fluorescent lamps) that the author obviously could not know, if not through the hypothetical extraterrestrial who brought them to Earth. The representations are usually very rough and those who propose these interpretations have insufficient scientific and technological knowledge. This problem has sometimes also been felt in traditional archaeology, where it may be difficult for an archaeologist to correctly interpret a technological object, but it becomes crucial when an extraterrestrial interpretation of a sketch or a sculpture is suggested. Think of the chances we have of correctly intepreting not only an object based on an alien technology, often beyond our scientific knowledge (already perhaps an

---

[4] Z. Sitchin, *The 12th Planet*, Avon Books, New York, 1978.

impossible thing in itself), but also the representation an Egyptian or Mayan artist might give of such an object. Any object set around a head may be interpreted as a helmet, and almost anything that is around a man may be a starship!

These theories would be no more than a harmless intellectual game if they were proposed and, above all, accepted with much healthy skepticism and considered for what they are—hypotheses, sometimes fascinating, but without any evidence or serious clue, ideas on which we can base, for instance, a fantastic tale, setting our imagination free. The movie and television series *Stargate* are a good example of it. The danger comes when the distinction between imagination and reality is lost and fantastic stories are mixed with the true facts of life.

Actually these fantasies, like many other pseudo-sciences, become extremely dangerous when they supply the basis for pseudo-religious cults, professed by sects with a strong propensity to fanaticism. The above-mentioned hypothesis, originated by the Sumerian mythologer Zecharia Sitchin, for instance, was taken up by the Raelian sect, founded by Claude Vorilon, who changed his name into Rael (from which the name of the sect) after an alleged encounter with an alien in 1973. The sect, like many others, has considerable wealth at its disposal, which allows it to use extensive technological means and to operate in many countries, exploiting loopholes in the laws and the possibility offered by the Internet to circulate these ideas in a quick and uncontrolled way. It seems that the Raelians' aim is to perform experiments of human clonation to imitate what the extraterrestrials did to create humankind. The sect founded a research company, Clonaid, and claims to be close to achieving its goal. Raelians are not alone and are perhaps not even the most dangerous; the danger of these sects must not be underestimated, as demonstrated by some bloody episodes in the past, such as the collective suicide of a group of youngsters who intended in this way to reach an alien spaceship orbiting in circumterrestrial space.

But even if these beliefs were not directly dangerous, they would constitute a more subtle danger, typical of all pseudo-sciences. The problem is not so much in the theories they advocate, but in their rejection (theorized or simply factual) of the scientific method and in the background of irrationality and fanaticism they imply. In the long run the conspiracy theory they assert on all occasions undermines society, and the distrust in science and technology makes solutions to the problems that our society must face more difficult. Other negative aspects are the reluctance of many scientists to take into consideration innovative hypotheses for fear of being confused with the supporters of these pseudo-sciences, and the difficulties in obtaining research funds from government agencies for disciplines like SETI. The reaction of the United States Congress to the appropriation of NASA funds for SETI is a recent example of this.

Besides the general lack of exceptional, or even convincing, evidence and the presence of simpler alternative explanations (Occam's razor), which, being more conventional, are less alluring and arouse less interest in public opinion, two interesting issues must be considered. The first is that very often the scientific statements that are supplied as evidence for these theories are outdated; the hypothetical alien science that was revealed to ancient Earthlings is very similar to the scientific theories that were in fashion in a recent past, often when the author of the pseudo-science attended high school, but that were later superseded by other theories of which he has no knowledge, since he is not professionally interested in scientific matters. The second is that the explanations proposed to solve one "mystery" are often incompatible with those advanced by another author (or even by the same author) to explain other enigmas. There is then no systematic explanation of a set of events, but many explanations contrasting with one another. Yet often these theories, even if inconsistent, end up reinforcing one another, since they have a common denominator based on the fact that official science (the government, the military, etc.) is hiding the truth. A typical example is the theory according to which Americans succeeded in landing on the Moon thanks to technologies coming from extraterrestrial space vehicles, or the theory that the whole landing on the Moon was a hoax. Clearly these theories are contradictory, but they end up strengthening the idea that public opinion is manipulated by the political and military establishments, and in this sense they support each other.

As a final consideration, these theories asserting that extraterrestrials landed on our planet in the past, interacting with our ancestors, playing the role of gods, and perhaps even creating gods, seem to pose a threat more to the established religions than to science. This is generally only marginally true.

Hindus, but also Buddhists and above all Jainists, for instance, tend to look with interest at all alleged discoveries suggesting that our civilization was preceded by other, more ancient, ones. Any discovery of an ancient high-technology object leading to theories in which humankind had in the past reached an advanced scientific and technological level, only to fall again into a primitive stage, looks like a confirmation of the cyclical view of history typical of Hindu scriptures. In this view of the world and of history, believers also look favorably on the idea that in the past there were contacts with extraterrestrials, who perhaps caused humankind to develop new technologies and civilizations. This is somewhat contradictory, however; if there were contacts of this kind, the cycle when this occurred must have been quite different from the others, and then the repetitive pattern that they have their faith in is broken.

However, religions based on the Bible are sometimes also unexpectedly open to these possibilities: in a recent public discussion between the above-mentioned Zecharia Sitchin and Father Corrado Balducci (reported on the Internet), the latter, a theologian with a significant following in the Vatican, admitted that the idea that humankind was the outcome of biology experiments by extraterrestrials is not unacceptable. Since ultimately these extraterrestrials must have been created by God, they could even be seen as a perhaps unintentional instrument of God.

Very likely Father Balducci is right; it is not up to theology or religion to assess whether these ideas may be true. Whether the presence of extraterrestrials in the past of humankind is a sound idea or not is up to science (all sciences are involved, from social and historical sciences to biology, from physics to astronomy) to say. And until now there is no doubt: these are just fantasies.

Yet, even if there is no evidence or serious clue, it cannot be ruled out that our planet has been visited by extraterrestrials in the past. Rather, the probability that a visit occurred sometime in the four billion years since our planet was formed is infinitely greater than that such a visit should happen during the lifetime of our generation. An occasional visit may have left extremely feeble traces. Despite the fact that one of the main worries of the designers of planetary exploration missions is not to contaminate the environments studied by automatic probes and astronauts, we have already scattered a considerable number of objects on the Moon and Mars, from scientific instruments to fragments of wrecked probes, from objects discarded by astronauts to the bottom part of the lunar excursion modules. On the Moon we even left a golf ball. Even admitting that aliens might have behaved like we did, it is extremely unlikely that any object of small size survived for a long time on the surface of an active planet like Earth.

But there is an additional difficulty: if an alien civilization arrived on our planet, it must have used a technology much more advanced than ours. How may we recognize objects produced by such a technology? It is possible that they used materials radically different from those we are accustomed to; metals, for instance, may be typical of a technology at our stage, and we already see the substitution of metals, on a growing scale, with plastics or composite materials. Many authorities foresee that microtechnologies and nanotechnologies will in the future allow us to build machine components and space vehicles with techniques more similar to those of biology than of mechanical constructions. That might explain why we are not able to recognize traces left by alien visitors: the materials they left could be, so to speak, biodegradable, and no trace of them remains or is detectable.

We must then consider every possible clue of such visits with an open mind and, knowing that the absence of evidence does not prove that such visits never occurred, we must state once more that to date no evidence, not even a single serious clue, of such visits has ever been found.

ALIEN PROBES IN THE SOLAR SYSTEM

But not only on Earth might we achieve contact with hypothetical extraterrestrial intelligences or with objects built by aliens. In addition, if an alien exploration team visited our planet in the past, it is reasonable to think that the visitors left one or more probes in the Solar System. Certainly, if a probe is left in the Solar System with the aim of transmitting data regarding a geologically very active planet like ours for long periods, it seems more reasonable to locate it in space, at a certain distance, where it cannot be damaged by volcanic eruptions, earthquakes, or other catastrophic events.

Therefore, if an extraterrestrial civilization studied Earth in the past, traces of its visits might be found in the neighborhood of our planet. A good observation point could be the near side of the Moon, a place where a probe can observe our planet without interruption, also keeping under control all possible transmissions we might broadcast (a kind of reverse SETI). Furthermore, the Moon is much less geologically active and the probe could survive for a long time, though the continuous bombardment by meteorites, and in particular by micrometeorites, could suggest that another solution be adopted. Alternatively, the probe might be located in orbit around Earth, at a suitable altitude, where it would be closer to the objects to keep under control and could thus perform its observation tasks more easily. If that task is to keep transmissions from Earth under control, it may also be located at a larger distance, depending on whether the hypothetical extraterrestrials want to keep its presence secret or not. One possibility is that the probe is located in the asteroid belt, where it would be almost indistinguishable from a small natural body, though it would be at risk of being destroyed by a collision, or even in the outer Solar System.

It has often been suggested that a radio relay station between two planetary systems could exploit the gravitational lens effect of the two stars. The Sun, like any other star, deflects light, radio waves, and other electromagnetic waves, owing to its large mass behaving like a glass lens in classical optics. Yet there is a difference: in optics, light converges to a single point, the focus, while a gravitational lens causes electromagnetic radiation to converge on a line, the focal line. Every point of this line is therefore like a focus of the lens.

The ideal position to locate a probe that must remain in radio contact with a distant extrasolar planet, therefore, would be a point on the focal line of the Sun's gravitational lens, passing through the latter and the star around which the extrasolar planet orbits. Since the focal line begins at a large distance from the Sun, more than 600 AU, instead of locating the probe at this large distance from the inner Solar System, the hypothetical aliens would perhaps only locate a radio relay station on the focal line, similar to our telecommunications satellites, to maintain contact between the probe, located closer to us, and their planet.

Recently, serious doubts were advanced on the actual possibility of using stars as gravitational lenses for interstellar communications, due to the mass asymmetries causing an effect similar to that of a defective lens and, above all, because of the enormous difficulties in keeping the transmitting and receiving stations always lined up with the two stars. The energy needed to maintain the alignment would likely be much greater than that needed to perform the transmission without exploiting the gravitational lens effect.[5]

The possibility, in the Solar System, of detecting alien probes whose aim is to study our planet and communicate with us depends on two factors: the size of the probe and the possibility that it broadcasts transmissions. If the probe is passive, or if its source of energy is no longer working, it may be extremely difficult to identify. If it were built with our present terrestrial technology, a scientific probe may have a mass from some hundreds of kilograms to a few tons and a size of a few meters. If the probe uses a radio link, the bulkiest part, and perhaps the most difficult to miniaturize, is the antenna, but its size might be more easily reduced if a laser is used to communicate with its base. Besides, our present technology does not permit the construction of a power system that guarantees continuous operation for centuries, let alone millennia.

---

[5] G. Genta and G. Vulpetti, "Some Considerations on Missions to the Solar Gravitational Lens," *Journal of the British Interplanetary Society,* vol. 55, no. 3/4, 2002, pp. 131–136.

Naturally, if we are looking for an alien probe, we should look for something far more technologically advanced, and here things become more complicated. It has already been predicted that, within a few decades, micro- and nanotechnologies will make it possible to build probes constituted by a single integrated circuit (probe on a chip) or at least of the size of a credit card. Above all, these microprobes would require less power than the present probes and, if aliens succeeded in solving the problem of the miniaturization of the antenna and power system, the whole spacecraft could indeed be microscopic. A very advanced technology might be able to build probes that are invisible to the eye—not larger than a speck of interstellar dust.

Clearly it is almost impossible to locate objects of this type in space, unless they can be found through their transmissions. Here, too, to save power, the probe could broadcast a very focused transmission, a simple thing in the case of a laser transmission, but much more difficult for radio transmissions. In this case the detection may occur only if the receiver is within the small zone encompassed by the beam.

After all, if alien probes are present in the Solar System, the only way we could detect them would be if they had been designed to be detectable, possibly even to contact us. Allen Tough holds that alien probes may be present in the Solar System and that currently they may be, so to speak, asleep, ready to be activated by a transmission from Earth. A probe of this type would monitor our transmissions and could have been programmed for answering a message showing that we have reached a given technological level. This approach is close to the *Galactic Internet* hypothesis seen earlier: the probe would be the node of the net closest to us and would only wait for us to apply to allow the connection.

Tough suggests that if a probe is monitoring our transmissions, it will pay particular attention to the Internet, since it is the best means of knowing everything about the present stage of our civilization. The simplest way to contact it would be to deposit a message in the Internet capable of activating the probe and inducing it to contact us. It may be objected that any message left in the Internet for this purpose will provoke an enormous number of answers, mostly terrestrial ones, and that it would be very hard, perhaps even impossible, to distinguish a possible authentic extraterrestrial message amidst the "noise" so generated!

An example of this strategy to establish contact with ETIs is the project *Invitation to ETI* started and coordinated by Allen Tough. Its Internet site[6]

---

[6] http://members.aol.com/WelcomeETI.

includes, among other things, a message to extraterrestrials with a warm invitation to answer ten questions on life and the Universe.

PLANETARY ARCHAEOLOGY

If intelligent beings visited the Solar System or if they inhabited it in the past, it is reasonable to expect that some traces of their passage might be found on some of the celestial bodies in the system. As mentioned earlier, it is likelier for these traces to still be present on less geologically and biologically active bodies, like the Moon or Mars, than on Earth.

Mars is the celestial body that was more studied from this point of view since the time when Schiaparelli, Lowell, and Flammarion observed the canals. In the 1960s, for instance, the astronomer Iosif Shklovskii explained the low density of Phobos, a satellite of Mars, by advancing a theory that it is just an empty shell, and is therefore an enormous artificial satellite. The astronomer then suggested that the ancient Martians, aware of the impending end of their civilization, created a huge orbiting museum, or library, to preserve its memory. At the time Carl Sagan gave much resonance to this theory in the West, with the purpose of suggesting ideas at the limits of what is reasonable, to encourage the imagination of the public and researchers and force the latter to perform every effort to disprove them. The confutation came with the first photos of the satellite taken by space probes (Figure 5.2): Phobos is without doubt a natural body, an asteroid captured by the planet.

Sagan's objective actually proved to be mostly a useless exercise. Theories like this are quickly and uncritically accepted by a part of the public, while researchers do not even bother to disprove them. A kind of incommunicability is so created: each side repeats its own truth without even listening to the reasons of the others, in a climate of suspect and mistrust.

Again, when the *Viking* probes started transmitting images of the surface of Mars in 1976, Sagan invited scientists to look for traces of archaeological interest (SETA: Search for ExtraTerrestrial Artifacts) on the Red Planet. It must be noted that at that time the hypothesis of the existence of living beings on Mars was more believable than it is today, even if the results of the previous probes had been very negative. The results of the *Viking* missions dealt the residual hopes a deadly blow, even if nowadays their critical revision is in progress.

Furthermore, the resolution of the images taken by the *Viking* orbiters

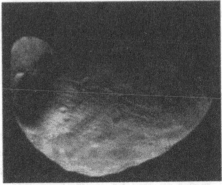

FIGURE 5.2 *Phobos, a satellite of Mars. In the 1960s a theory according to which the satellite was artificial, a big orbiting library built by a Martian civilization, was proposed. Actually, as the probes showed, it is an asteroid captured by Mars. (NASA image.)*

was too low to reveal traces of intelligent life, unless they were in the form of huge, monumental constructions. The result of that search was negative, at least as far as the official data are concerned. Yet an image taken in the Cydonia region revealed an object that many interpreted as an enormous human face, a kind of sculpture of the type of those on Mount Rushmore, but larger. This imaginary sculpture has been given the name of "Cydonia Sphinx" (Figure 5.3). That region on Mars was not photographed again until 1998—that is, 20 years later—and with no new data the imagination of a part of the public was free to give the most fanciful explanations of this rock formation.

These interpretations opened up an infinity of arguments. The discovery of an artificial object would in itself be a result of paramount importance, posing endless further questions, such as who built it—a civilization that developed on Mars or visitors coming from a distant planet? In the first case we should start from Schiaparelli's and Lowell's dreams again, but such a perspective seems to be in contrast with what we know of the Martian environment. In particular, those who defend this thesis start from the assumption, much accredited nowadays, that the planet had a climate in the past very different from the present one, but they postpone by billions of years the series of events that brought it to its present state.

If the planet cooled three or four billion years ago, in geological terms shortly after its formation, there would have been no time for life to evolve to the stage of intelligence, and it would be almost impossible for such an ancient structure to survive to the present, even on a planet like Mars. We must not forget that in the past its surface was not as quiet as it is now. To overcome this objection, it has been suggested, also in light of the discoveries on the effects of meteoric impacts on the terrestrial

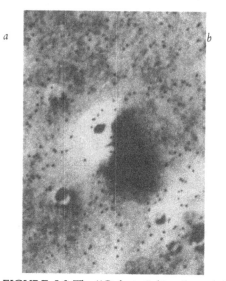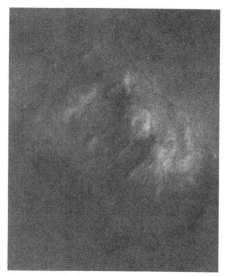

FIGURE 5.3 The "Cydonia Sphinx," a rock formation in the Cydonia region on Mars. Many believe it is a monumental structure built by aliens. (a) Photo taken by the Viking 1 orbiter on July 25, 1976; (b) photo taken by the Mars Orbiting Camera (MOC) of the Mars Global Surveyor on April 5, 1998. (NASA images.)

environment, that Mars was actually similar to Earth until some hundreds of millions of years ago, when the impact of a meteorite dispersed the atmosphere, causing the evaporation of the seas, cooling the planet, and completely destroying all life. Some think that such an impact occurred at the same time as the K/T extinction on Earth; a body belonging to the same swarm of meteorites as the one that caused the extinction of the dinosaurs reduced Mars to its present state.

In this case there would have been enough time for an intelligent species to develop a civilization, and the age of the Cydonia Sphinx would be more compatible with its survival to the present time. Others have even suggested that the cooling of the planet actually occurred a few thousand years ago, including the end of Martian civilization in a hypothetical catastrophe—the explosion of a planet orbiting between Mars and Jupiter—that affected the entire Solar System. This catastrophe, which would have occurred no earlier than 10,000 BC, being therefore almost contemporary with us in geological terms, would also have produced the asteroid belt. This theory has many followers of various kinds, and fits very well within the theories trying to assert that the history of the Universe is much shorter than science has ascertained. It fits particularly well with creationist beliefs, for which there was no evolution but only catastrophic events, the last one being the Universal Flood.

No convincing evidence of such theories has ever been produced, and the very concept of the explosion of a planet has no scientific basis, while the presence of the asteroid belt is well explained by the failure of a planet to form because of the perturbations induced by Jupiter.

Another aspect of the Cydonia Sphinx attracted much attention: its strong anthropomorphic aspect and a certain (more alleged than actual) similarity with the Sphinx of Giza, the true one, built by ancient Egyptians. Here the fantastic interpretations are many, to the point that the human aspect is ascribed to the fact that the Martians abandoned their planet to move to Earth. Of course, none of those who assert these reveries was ever able to produce the slightest evidence of them, not to speak of the exceptional evidence needed to confirm such an exceptional discovery. After 20 years of speculation, the images taken by the *Mars Global Observer* caused a healthy return to the real world, confirming what NASA had always asserted: the "statue" is just a natural formation that, in particular light conditions, gave the impression of an artificial structure.

Needless to say, the most ardent supporters of the artificial origin of the formation did not change their mind, but rather accused NASA of hiding a disturbing truth. On this matter a passage by Graham Hancock can be quoted: "Like other big state bureaucracies, NASA has lied and will lie again. We think the evidence suggests it has lied about Cydonia ever since the face on Mars was discovered."[7] He goes to the point of asserting that Carl Sagan also knew the truth when he wrote that the face is an optical illusion, since he knew both the true pictures (which ones?) and those that were modified, and was therefore part of the conspiracy. It is quite easy to accuse a person who is deceased and can no longer defend himself. However, the delirium continues with words that are worth quoting: "Cydonia is indeed some sort of signal—not a radio broadcast intended for an entire universe, but a specific directional beacon transmitting a message exclusively for mankind." The aim of its authors is to tell us that there is a dangerous asteroid aimed at Earth and they built the face in such a way that it points toward it. If we are intelligent enough (and obviously if NASA will stop deceiving us), we will be able to detect the asteroid and destroy it before Earth is reduced to a desert like Mars.

If all of this is unbelievable, this last statement is really foolish; as the famous mathematician Jules-Henri Poincaré realized at the beginning of the twentieth century, and the studies on the theory of chaos and nonlinear

---

[7] G. Hancock, *The Mars Mystery: The Secret Connection Between Earth and the Red Planet*, Crown, New York, 1998.

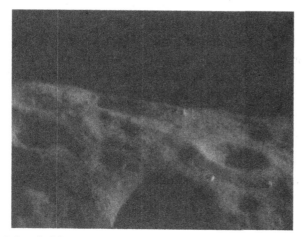

FIGURE 5.4 *Detail of the surface of the asteroid Eros, in an image taken on August 6, 2000, by the probe* NEAR-Shoemaker *from a height of 49 km. Someone interpreted the reflecting zones as artificial structures. (NASA image.)*

astrodynamics have now demonstrated, celestial dynamics, and particularly the dynamics of small bodies moving under the action of the combined attraction of the Sun and of one or more planets, is not predictable as was thought in the past. The trajectory of an asteroid, and particularly of an asteroid whose orbits intersect that of a planet, cannot be computed years (or millennia, in this case) in advance with such precision as to state whether it will collide with a certain planet. And this not for the inability to calculate the trajectory or for lack of suitable instruments, but for precise theoretical reasons. No alien civilization, even much more advanced than us or endowed with very powerful instruments (unless it is a civilization of magicians with suitable crystal balls) could ever tell us that a certain asteroid will bring havoc upon our planet. They could warn us that a certain asteroid has a certain probability of striking us sooner or later, but this we already know: large asteroids have already hit Earth and others will continue to do so. As usual, the science of the aliens finds its limitation in the scientific knowledge of those who invented them, who are seldom experts in the disciplines involved!

It is interesting to note that the scientists who conducted the search for life on Mars with the *Viking* probes were accused of having preconceived ideas and of refusing to find traces of life by performing experiments badly suited to this task. NASA is accused of keeping the evidence of extraterrestrial life secret. At the same time, it is accused of trying to endorse at all costs the presence of traces of life in the Martian meteorites to sway public opinion to demand that Congress approve huge funding for a human mission to Mars. Undoubtedly, no incentive for an expedition to

Mars could be better than the discovery of an artificial structure. Why would NASA keep something like the Sphinx secret?

But it is not only on Mars that artificial structures are alleged to have been found; antennas, domes, and other traces of buildings are said to have been photographed on asteroids (in particular on Eros, Figure 5.4). Again there is no evidence for such statements.

Artificial objects on asteroids would indicate the presence, now or in the past, of an intelligent species able to travel through space in our Solar System, since it is extremely unlikely that life could evolve on an asteroid. The origin of such a species would quite probably be extrasolar and would therefore constitute an answer to the Fermi question.

INTERSTELLAR PROBES

If, as Fermi would say, we do not see any aliens in the Solar System, some objects built by humankind are slowly entering the closest zones of interstellar space. Their speed is very low, but in the vacuum of space they may remain intact for millions of years, traveling for distances of many light-years.

Even if it is unlikely that sooner or later someone finds a probe launched by humans of planet Earth in the immense expanses of interstellar space, such an occurrence cannot be completely dismissed. For this reason the *Pioneer* probes carry an engraved plate with some sketches that should give an idea of the origin of the probe and of the beings who built it (Figure 5.5).

A more complex message was prepared for the *Voyager* probes; they carry a phonographic disk on which a selection of sounds and images of Earth is recorded, in the hope that if aliens ever find it, they will be able to decode the message.

The choice of what to include in such a selection was very difficult. There was not only a limitation of available space on the disk, but the information had to be presented in a way that might be understandable to intelligent beings of whom we know nothing, and a picture of our civilization had to be given in an unbiased way. Moreover, the whole job had to be performed in a short time, while fighting against bureaucratic difficulties of many kinds.[8]

---

[8] An interesting description of the content of such disks and of their preparation is reported in C. Sagan, F.D. Drake, T. Ferris, J. Lomberg, and L. Salzman Sagan, *Murmurs of the Earth*, Random House, New York, 1978.

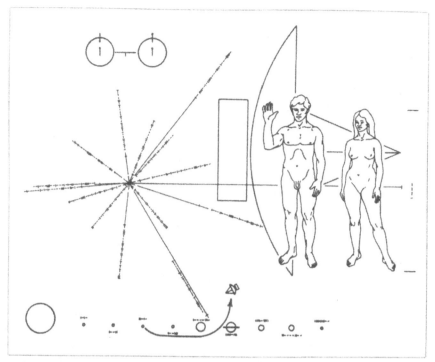

FIGURE 5.5 *Plate carried by the* Pioneer 10 *probe. From the top in a counterclockwise direction: sketch of a hydrogen molecule; map of the pulsars with information on the launching place and time of the probe; schematic sketch of the Solar System with an indication of Earth's position and of the first part of the trajectory of the probe; sketch of two human figures near the antenna of the probe that gives the scale of the sketch. (NASA image.)*

The disk, a 33-rpm phonographic disk (also recorded to work at 16 rpm), made of gold-plated metal instead of plastic material and protected by an aluminum case (Figure 5.6), has a sketch describing the procedures needed for playing it back engraved at its center, where the label is located in ordinary disks. Together with the disk, the probes also carry a pickup, to facilitate the task of the extraterrestrial who might find it.

The selection of the color photos, black and white photos, and sounds and music is intended to give an idea of our planet and, above all, of human civilization. An initial choice was to exclude images of wars, natural catastrophes, and human violence, and some difficulties were encountered with the images regarding human anatomy. NASA had already received much criticisms for the *Pioneer* engravings on Figure 5.5 (from the charge of sending around in the Universe obscene images to that of antifeminism because, while the male figure is greeting, the female one has an attitude that was interpreted by someone as subdued) and the space agency did not want to risk further criticism.

As this will reemerge each time an interstellar probe is planned, perhaps it is worth mentioning some of the problems in representing human anatomy in graphic messages like those of the *Pioneer* and *Voyager* probes. The plate of the *Pioneer* probes initiated many discussions and jokes, particularly in the United States, showing that the taboo of nudity is still very strong. Perhaps the most amusing cartoon is that in which two aliens, very humanoid and completely covered, look at the probe, and one says to the other: "So the Earthlings are like us, with the only difference that they do not use clothing!" Apart from the obvious observation that immediate conclusions should not be drawn from any extraterrestrial message, this cartoon can make us think about a, perhaps marginal, but interesting fact. If intelligence and technology are so linked to each other—whoever finds a probe must have an advanced technology, since he must look for it in space, and if a *Pioneer* ever crashes onto a planet the plate would be so damaged that it would not be possible to read it—and if an intelligent species tends to spread over much of its planet, it must use clothes as a protection against adverse climatic conditions. Clothing is one of the first technologies humans used to live, where a fur like that of a bear is preferable to the short hair of a leopard. Once the skin is protected from the cold, any natural fur would be of little use and evolution would probably eliminate it. An intelligent species deriving from the primates will therefore be a species of "naked apes" (to paraphrase the title of a well-known book) and, similarly, other intelligent species will be naked but artificially clothed. It is unlikely that we will meet an intelligent being like Chewbacca, one of the main characters of *Star Wars*, covered by a bear- like fur. If clothing is a distinctive sign of intelligence, the taboo of nudity might also be widespread, and therefore it is possible that images like those of the *Pioneer* might perhaps disturb not only the more narrow-minded inhabitants of Earth, but also the aliens who receive the message.

As already said, it is likely that this is not a problem at all, since some of the transmissions we are broadcasting violate the taboo of nudity (and quite a lot of other taboos too) in a much more explicit way, and at any rate if we send a message with the aim of being understood, we should not bother too much about the reactions of others (also because we cannot know what they are) and give as straight and open a representation of ourselves as possible. It might also be important to explain how humans and all the complex beings living on Earth reproduce; any sketch related to this topic will be for them an almost obvious confirmation if their evolution has chosen a similar path (sexual reproduction with two separate sexes); it will be an interesting discovery if their method differs

FIGURE 5.6 Outer case of the disks carried by the Voyager *probes. (NASA image.)*

from ours, but not too substantially (for instance, sexual reproduction with more than two sexes, parthenogenesis, etc.); or will be something impossible to understand if they reproduce in a radically different way (budding, or ways we totally ignore).

Another choice was not to send images of religious scenes to ensure that some religions are not given prominence with respect to others; the only building somehow linked to religion is the Taj Mahal, included not as a temple but as a monument to love. On the other hand, it was thought that the human religious soul could be imagined from some music, such as that by Bach.

From the disk the will of representing the best of humankind is evident, and, besides, scenes of violence could be interpreted as a threat toward other species. An image with a man who embraces the galaxy was rejected, since it might be interpreted as an hegemonic statement. Someone observed that the disk represents more how humans would like to look to hypothetical extraterrestrials than what they really are.

The disk also contains greetings in 54 languages, including some that are no longer spoken, such as ancient Sumerian. The recorded sentences go from simple and colloquial greetings to solemn phrases like the Latin *Salvete quicumque estis; bonam ergo vos voluntatem habemus, et pacem per astra ferimus* (Hi, whoever you are. We come to you in good will, and we bring peace among the stars). The song of whales was also recorded together with human voices.

The then-General Secretary of the United Nations, Kurt Waldheim, and the then-president of the United States, Jimmy Carter, wanted to include an address, making the *Voyager* a first tool of galactic diplomacy.

The message from Waldheim was as following:

As the Secretary General of the United Nations, an organization of 147 member states who represent almost all of the human inhabitants of the planet Earth, I send greetings on behalf of the people of our planet. We step out of our Solar System into the Universe seeking only peace and friendship, to teach if we are called upon, to be taught if we are fortunate. We know full well that our planet and all its inhabitants are but a small part of the immense Universe that surrounds us and it is with humility and hope that we take this step.

The message from Carter ended with these words:

This is a present from a small distant world, a token of our sounds, our science, our images, our music, our thoughts, and our feelings. We are attempting to survive our time, so we may live into yours. We hope someday, having solved the problems we face, to join a community of galactic civilizations. This record represents our hope and our determination, and our goodwill in a vast and awesome Universe.

The engraved plates of the *Pioneer* and the disks of the *Voyager* are therefore "bottles with a message" launched in the immense ocean of the cosmos, in the hope that someone will some day be able to find them, learning that a form of intelligent life exists on planet Earth. Or better, that it existed, since the probe cannot be recovered before some millennia, and it is very likely that either humankind will by then have expanded in space (and in such case it will not be necessary to look for messages engraved on the probes to contact it), or will be extinct.

If it is highly unlikely that the probes that transport these message will enter the inner zones of some planetary system, it is not impossible that they will be found by a civilization that has developed interstellar travel and, possibly, has the ability to search for interstellar ships adrift in space.

If within some thousands years humankind will have similar abilities, it is not unlikely that the messages we launched in space will be recovered by our descendants, supplying them with archaeological material that enables them to understand our civilization. The idea of launching documents for our descendants has been considered several times and was also put in practice by including some documents in the satellite LAGEOS, whose orbit will allow it to remain in space for several million years.

THE SPACE IMPERATIVE AND HUMAN EXPANSION IN SPACE

In the previous chapter the possible conscious life expansion principle was mentioned. It states that every biosphere, or at least every biosphere that succeeded in producing an intelligent species, tends to expand in space.

Humans of Earth are now trying their first, often uncertain, steps, in the Solar System. As in all beginnings, the difficulties are enormous, since for every step that is made to leave our planet, all the materials and people must be transported from the gravitational well of Earth.

The costs of any scientific or commercial space mission is strongly influenced by the launch costs and are therefore extremely high. From this point of view, only a drastic reduction in launch costs may induce an acceleration of space exploration and exploitation.

Yet, many of these difficulties are bound to decrease, thanks not only to new technological developments, but also to our increased activities in space. It is obvious, for instance, that the cost of installing the first lunar outpost will be enormous, but the cost of the following ones will be much smaller as most of the material needed will be extracted from the Moon itself. Another example is that of the large space habitats: it is almost impossible to build them if we have to carry all the materials into orbit from Earth, but the existence of a lunar base and of plants for the extraction of materials from our satellite would make them feasible at a more reasonable cost.

The beginning of the transformation of humankind from a species living on a single planet to a spacefaring civilization, diffused at least in a whole planetary system, will be slow, even very slow (the time needed will depend on many factors, some technological but mostly financial), yet the accumulation of technologies and infrastructures will make it inevitable and probably irreversible.

Actually, while remaining in the Solar System it will be difficult for humankind to make contact with extraterrestrial intelligences, unless it can do so indirectly. It is clearly much more possible to find a probe or another alien object during the exploration of a celestial body of the Solar System than to find one on our planet and, as previously stated, it is easier to discover extraterrestrial forms of life, or to receive artificial transmissions using large observatories on the Moon or in space, than using telescopes and antennas located on the surface of Earth. These instruments are located at a short distance from our planet (with the exception of a possible use of the gravitational lens of the Sun for SETI, an extremely unlikely occurrence), but probably only a species that has started a large-scale colonization of its own system has the technological ability to install large and complex research equipment in space.

The actual possibility of physical contact between different intelligent species begins the moment one of them is able to travel beyond its system to explore interstellar space.

As discussed in Chapter 4, there is no physical reason why this event should be impossible, at least as far as long journeys taken at speeds less than the speed of light, or even at a relativistic speed, are concerned. In the latter case the trip, which will be of very long duration for those who remain at home, may be much shorter for those who travel. The objections to this type of interstellar travel are, above all, financial, owing to the enormous quantity of energy required.

Such objections, however, seem poorly motivated. Certainly the high cost of interstellar travel makes it impossible for a society at our present level of development—and, in general, for a society whose energy resources are limited to those available on a single planet—to colonize other systems. But a civilization that is able to exploit the energy resources of its own planetary system can easily circumvent this energy problem; interstellar journeys will probably remain a very expensive activity, but they will not be impossible. At present, a space ark or a starship is no more distant from our technological abilities than a nuclear aircraft carrier was for Phoenician shipbuilders!

It has, however, been noted that interstellar travel will be restricted to journeys of colonization, since, by definition, colonization requires only one-way trips. Humankind (and the other intelligent species, if the conscious life expansion principle really holds for every species) can then spread into a more or less larger zone of the galaxy. The difficulty of two-way journeys—which may be possible if relativistic speeds are reached, but at any rate are unlikely—relegates interstellar commerce and the creation of political entities, including several systems, to the realm of dreams. The galactic empires, imagined first by Isaac Asimov and then made popular by the *Star Wars* saga, will always, therefore, remain beyond practical possibilities. And without such large political entities, not only will interstellar warfare not occur, but there can also be no humanitarian assistance should a community in a distant system be involved in a planetary catastrophe.

Owing to the vast distances involved, each inhabited system would therefore be a kind of island, maintaining contact only through radio (or optical) broadcasts with other systems. Besides, this contact would be only sporadic and slow. We could not even hold an actual dialogue, since many years would pass between a communication and the reply, and this could expand to centuries or millennia in systems at greater distances from each other. The information exchange would be more aimed at informing the various communities of the history and science of the others, but we cannot think that such slow exchanges will allow humankind to maintain a cultural unity. If today we are at the beginning of a global civilization at the level of species, and if such substantial uniformity can also be maintained when the human species begin to spread in the Solar System, then interstellar expansion will cause new and significant differentiations.

These differentiations may go well beyond the cultural aspect, and change our species even at a somatic level. The evolutionary pressures of different environments and the attempts aimed at adapting humans to the various environments through genetic engineering may perhaps differentiate the human species still further, to the point of producing completely different species.

Any possible contact with extraterrestrial intelligence must be considered in a scenario of this type. From a certain point of view, it may just be a future part of our own human species that makes contact with these intelligences. People living in systems far from the one involved would receive information only after a long time lag, thanks to radio transmissions, and for them it will not be much different from what could be today the discovery made through SETI research. Yet, if there is physical contact by some humans, perhaps it will be much easier to reach some form of understanding. In fact it will be far easier to communicate with beings with whom an almost immediate interaction is possible, making several attempts at questions and answers, than with distant beings, where every attempt at interaction requires years or perhaps centuries. Perhaps physical contact (to be able to see and touch them) is not so important from this point of view, while it would help to be at a distance that allowed an almost immediate exchange of information. A distance equal to that between the Earth and the Moon, therefore, could be considered "optimal."

Besides, we must not forget that the different environmental requirements of the various species may not allow the simultaneous presence of all representatives in the same place, and therefore a true close contact may not be possible.

HOSTILE CONTACTS

One of the classical themes of science fiction is the invasion of Earth by extraterrestrials or the war between species trying to colonize the same planet. It is clear that situations of this type are very promising from the narrative point of view because they take the conflict to its maximum level, which is the key of any fiction, but do they represent a true danger that we must take seriously into consideration, or do they involve only highly unlikely scenarios?

The possibility of interstellar travel, making it possible to have physical contact between intelligent species, does not allow for the total exclusion of the contact being hostile. In particular, if the conscious life expansion principle is realistic, the reason for which interstellar journeys are performed is colonization, and this by itself is a possible source of conflict. On Earth, when a species reaches a zone where the ecological niche it tries to enter is already occupied, a conflict, not a direct one but in the form of competition for the same sources of subsistence, ensues. From this conflict a winner emerges, in the form of the fittest to the environment. This happens in the case of nonintelligent species; however, the presence of intelligence should change this behavior completely, but the only precedent we have is what happened each time an expanding human community entered into contact with another community that already occupied the territory under consideration.

The most often mentioned examples are those of the colonization of America, and of European colonialism of the nineteenth century, in which the native populations succumbed, to the point of cultural assimilation and, in some cases, of a true genocide. But it is not something peculiar to European society: the peoples that came in contact with another, technologically more advanced or numerically more powerful, civilization in other places and other times have sometimes met an even worse fate. Jared Diamond quotes some illuminating examples in *Guns, Germs and Steel*.[9] The destruction of the Morioris by the Maoris, perpetrated in 1835, is worth considering because it is a well-documented episode that, affecting small populations and lasting only a very short time, can be described in detail. The Maoris colonized New Zealand about 1000 BC; then a small group went back to the sea and traveled eastward, discovered the Chathlam Islands and settled there. The two communities began a completely independent development, without any contact; the Maoris developed a very organized and warlike society, with a strong technological development, while the Morioris, as the inhabitants of the Chathlam Islands were called, remained essentially gatherers, with a low technological level and a pacific and a little organized society.

On November 19, 1835, a ship with 500 armed Maoris reached the

---

[9] Jared Diamond, *Guns, Germs and Steel*, W.W. Norton, New York, 1999.

coast of one of the islands, followed on December 5 by a second ship with a further 400 warriors. The Maoris summoned the inhabitants of all the villages to submit themselves as slaves and the Morioris, instead of trying to fight (they actually were more than twice as many as the invaders), accepted. Nevertheless, the Maoris attacked and in a few days killed almost all the natives—men, women and children—and ate them. The few who were not killed were considered as slaves, but were killed and eaten subsequently. The words of one of the Maoris are meaningful:

> We have taken possession of the island, according to our customs, and we have captured everybody. Nobody succeeded in escaping. We killed those who ran away, and so all the others. But what does it matter? These are our customs.

Diamond notes that this brutal outcome could have been easily predicted:

> It is natural that when two so different peoples come into contact it is the first one [the less organized and technologically advanced] to succumb and not vice versa.

This scenario reminds us of so many alien invasions described in science fiction, from *The War of the Worlds* by H.G. Wells to the movie *Independence Day*, obviously except for the ending: the invaders do not attribute any value to the lives of the natives and they only want to annihilate them to obtain what they want, usually the natives' land.

Today we think that behavior of that kind is barbaric, and we say that if we were to enter the territory of intelligent beings (from this point of view it is immaterial whether they are humans of Earth or aliens) we would behave much differently, and we are probably sincere. Nevertheless, perhaps it is not very correct to think that modern humans have now overcome this barbarian behavior, since half a century ago circumstances not better than those mentioned above (apart from eating the defeated people) occurred in "civilized" Europe. If we stop for a moment to think, we cannot blame any possible extraterrestrials for being very careful not to reveal their presence to us (another version of the zoo hypothesis, perhaps the least flattering for us).

The idea forwarded by Drake, that a civilization arriving at a stage of technological progress to travel in space must necessarily be beyond the phase in which it has an aggressive behavior, is convincing but only to a point, because of the anthropomorphism it implies and for the generalization to many species that may differ from each other. But a discussion on this aspect and on the likely age of the species with which we may come into contact, and on the duration of their life, was included in Chapter 4.

Here it is relevant to note that interstellar distances are in themselves a very strong defense against possible hostile contacts, and the fact that Earth was never occupied by extraterrestrials, neither friendly nor hostile, in its long history, is reassuring—that is, if we exclude the idea that the extraterrestrial invaders of this planet were our ancestors, as many seem to believe!

Consider a scenario where two intelligent species, able to travel between the stars of our galaxy, approach dangerously close to each other in their expansion. One of the species might be ourselves. It is not thinkable that before launching a colonization expedition one of them does not carefully study the destination planet and does not realize that the system is occupied by the other species. At this point it is reasonable that the two species try to reach an agreement and establish peaceful relationships, turning their expansion toward other directions. This may generally be due to reasons that may be defined as "ethical," linked with motivations of the type of those explained by Drake, or simply to reasons linked with their best interests. In fact, a colonization expedition at interstellar distances, whether it is performed using space arks, hibernated crews on small starships, or other solutions that only the aliens know, is extremely expensive and does not allow for the transportation of a large quantity of people and machines. Except in the case of an enormous technological superiority, logistic difficulties should discourage hostile intentions. In interstellar strategy, the defenders should enjoy such advantages to discourage any potential attackers.

It is necessary to say expressly that space colonization will probably never be seen by any intelligent species in the same way as the population colonies on Earth. The cost and duration of the trip are such as to make it unthinkable to transfer a large number of persons at interplanetary, and still more, interstellar distances, as a method of reducing any demographic pressure on the planet of origin. On Earth, demographic pressure has always been, particularly in ancient times, the main cause for mass migrations and invasions; this cause will very likely be absent in space colonization, by any species. Without this mighty incentive, the tendency to generate conflicts should be reduced.

The only true danger could come from Von Neumann machines, provided that they are possible. Actually, anyone who has autoreplicating automata at his disposal to perform interstellar travels could deploy, at an extremely limited cost, such a military power to be temped to try to solve problems by using force. But perhaps, as Carl Sagan realized, Von Neumann machines constitute a danger of even greater type, to the point that, in his answer to the proposal by Frank Tipler of using Von Neumann probes for interstellar colonization, he wrote:

> The advisable line for a technological civilization is that of banning the construction of interstellar Von Neumann machines and strictly limiting their use. If the argument of Tipler is accepted, such an invention would jeopardize the whole Universe; the control and the destruction of interstellar Von Neumann machines will become a task with which all civilized countries—the more technologically advanced, in particular—will in some way be involved.[10]

Tipler counters this thesis and asserts that if humankind gives up the opportunity of using Von Neumann machines, it will lose all possibilities of colonizing, firstly, the systems closest to us and then the Universe, betraying as a consequence its cosmic vocation and condemning itself to extinction, with serious consequences of physical and metaphysical order. He states:

> That reasoning is dictated by fear and ignorance, by a definition by exclusion: what is different from myself is not worth existing. A "person" is defined on the basis of the qualities of the mind and the soul, and not on a particular body form. Adolf Hitler didn't agree with this definition. ... If the philosophical base of racism is rejected, then such refusal must be accepted with all consequences: any law limiting the creation and the reproduction of intelligent machines must be opposed. At the end the intelligent machines will become more intelligent than the members of the Homo Sapiens species and will dominate civilization.[11]

Yet to extend the concept of a person to Von Neumann probes—even if it is consistent with Tipler's vision according to which a human person is only a particular software running on the hardware constituted by the brain—is surely far-fetched and one can think that Von Neumann machines are after all only machines, and moreover dangerous ones, without being either racist or Nazi!

The danger inherent to self-replicating automata lies in the fact that it is impossible to guarantee that, after a certain number of replications, the hardware but, above all, the software of these machines will continue to follow the original design. Errors and modifications will certainly be present, exactly as happens with the mutations of living organisms, and therefore there will be an evolution in an unpredictable direction that will probably cause the machines to adapt to the environment and pursue ends

---

[10] C. Sagan, quoted in F.J. Tipler, *The Physics of Immortality*, Mondadori, Milano, 1994, p. 84.
[11] Ibid.

that can be different from those of their original builders. As soon as the machines expand in space, the distance, in space and time, from their builders will increase and the latter will be less and less able to control the situation; if we launch a Von Neumann probe toward Alpha Centauri we could check surely by radio the program of the probes of the following generation, but only at the cost of a long delay in the prosecution of the expansion and with a nonnegligible energy expense for the transmission of the data. But the ensuing replications, which would happen at increasing distances, would be completely beyond our control.

In a scenario of this type, the behavior of self-replicating machines could become dangerous for humankind itself, but especially for the species they could meet in their expansion. More than directly from other intelligent species, the danger could come from automata that the civilizations that did not follow the suggestion by Sagan could have scattered in our galaxy. Luckily there is no evidence that automata of this type can actually be built.

In conclusion, it may be stated that the enormous interstellar distances and the even greater distances separating the systems where life may have developed makes the dangers linked with contact between species quite negligible after all. If "not to give away our position" may be a prudent measure, not unreasonable after all, we must nevertheless recognize that we have already violated it and we continue doing so with omnidirectional broadcasts that are clearly artificial and have enough power to be received from a distance of hundreds of light-years. A few transmissions done using the radiotelescopes, very focused and of short duration and therefore hardly detectable, or some probes that will remain for millennia close to the Solar System, will not make us take serious risks.

FASTER THAN LIGHT INTERSTELLAR JOURNEYS

The situation described in the previous section, although reassuring, might seem after all disappointing. Science fiction accustomed us to something much more exciting: starships rapidly crossing our galaxy in all directions, humans exploring unknown planets one after the other, aliens living elbow to elbow with humans of Earth and showing the same general virtues and vices. Are these just flights of the imagination or is there some possibility that these scenarios, with their thrilling perspectives but also their dangers, might become reality in a very distant future?

At first the answer must be negative and not only for reasons linked with the present state of our technology, such as the cost of interstellar travel or the large quantities of energy they require, but for an actual, scientifically demonstrated, impossibility.

The theory of relativity is clear on this: no material object or transmission of information may move with a speed that is faster than light (FTL). The theory of relativity was confirmed by countless experiments and today very few physicists have doubts about it.

Science-fiction writers tried to circumvent this obstacle by inventing a number of tricks to reconcile the hard scientific facts with their dreams and inspiration. In general, they tried to avoid what is an actual cosmic speed limit by using another result of modern science: the awareness that the Universe is much more complex than it seems and, besides the three space dimensions we are aware of and the time dimension, other dimensions that are beyond our comprehension may exist. Would it not therefore be possible to warp space–time in such a way as to cause the points of departure and arrival of the journey to be closer to each other, so shortening the way and reducing the time?

Are these just fantasies or is there some hope that this might be possible? Undoubtedly we must not think that modern science represents the limits of human knowledge; moreover, there are without doubt many things that science is not able to explain. The same theory of relativity might in the future be superseded by other scientific theories that will include it (since it was validated by many experiments), in the way that Newtonian mechanics is included in relativity as a particular case. A deeper knowledge of the Universe may perhaps lead us to perspectives that today are unimaginable.

On the other hand, general relativity has not yet been studied in enough depth, and, for instance, some solutions of its basic equations yield particular types of black holes, often called *wormholes,* that look impressively like the space–time tunnels of science fiction.

Today scientists are more open than even in the recent past, and NASA itself started the BPP (Breakthrough Propulsion Physics) program within the frame of the Advanced Space Transportation Plan, whose aim is to investigate the possibility of innovative concepts in the field of space propulsion. As the slogan of the project says, its purpose is *to make believable progress toward unbelievable perspectives.* Since the project is still based on very hypothetical concepts, it is focused more on the physical and mathematical aspects than on technology. The purpose is more to formulate the bases on which new technologies might be developed than to work on applied projects.

Wide-ranging research must be performed on a number of still unclear phenomena, trying to identify the concepts that may yield results in shorter times and with acceptable costs. The accent on the costs is linked with the fact that this is a high-risk research, in the sense that many different approaches must be attempted and many of them will prove to be unfruitful or at least to lead to inconclusive results. It is not therefore justifiable to make large investments on single research lines that could be in vain or, at least, too premature.

Great efforts will be directed at attempts to develop propulsion devices not based on the principle of reaction propulsion, allowing high thrusts to be obtained without the expulsion of matter, trying, for instance, to use coupling among gravitation, electromagnetism, and space–time, as described by general relativity. Today the accepted theories foresee that to manipulate inertia, gravitation, or space–time, electromagnetic fields larger by many orders of magnitude than those obtainable with our present technology, or those predicted in the future, are needed. Potentially, other promising phenomena are those linked with quantum fluctuations in a vacuum and with the hypothetical interactions between a gravitational field and rotating superconductors in magnetic fields quickly varying in time. Even if these phenomena are potential candidates for the realization of radically innovative propulsion systems, they are still hypothetical and must be first understood and interpreted in a consistent theory. Only after that can their technological exploitation be attempted.

It is recognized that propellantless propulsion devices could allow very high speeds to be attained, but they would always be less than the speed of light. As already stated, this limit comes directly from the theory of special relativity, while that of general relativity seems to offer some possibilities to avoid it. The so-called wormholes are particularly interesting not only because they would allow us to connect very distant points with a path that is much shorter than their space distance, and allow us to travel many light-years in a short time, but also because the gravitational effects within them do not seem to be particularly intense— which would allow material objects to cross them. For now they are only possible solutions to some equations and their actual existence has still to be verified.

In 1994 Miguel Alcubierre suggested that a warping of space–time might be created to allow a spaceship, moving inside it at a speed lower than that of light, to fly along a path in a time shorter than that taken by light. A kind of warp drive might so be obtained. It seems that the Alcubierre warp drive has been proven to be physically unattainable, but work on similar concepts goes on. There are many other ideas on which research work is proceeding, like the variation of inertia, the use of hypothetical particles that travel at superluminal speed (the so-called tachions), and the much hypothetical negative mass of matter.

Any direct attempt to go beyond the cosmic speed limit must face the fact that, at least conceptually, the motion at superluminal speed is connected intrinsically with time travel, with all the paradoxes and conceptual difficulties that the latter involves.

Even if not tightly linked with interstellar travel, another classical theme of science fiction is teleportation,[12] which could allow objects and people to move at the speed of light. Actually up to a few years ago it was completely ruled out by physics, since it entails many problems and incongruities. The idea on which it is based is that, as all atoms of the same type are exactly identical, it might be possible to extract all the information related to the structure of an object, to broadcast it by radio, and then, using other atoms, to reconstruct in its place of destination an object that is absolutely indistinguishable from the first one.

The first problem is what happens if the original object is not destroyed, or if the information is used to produce more copies of the same object. If it is an inanimate object, it does not produce serious philosophical consequences; rather, it would be the final solution to the problem of mass production. Note that the information could be originated by a computer making copies of virtual objects—a sophisticated rapid prototyping machine. But if a living being or, even more significantly, a conscious being is used, what do the copies mean? Does the "self" of the subject transmigrate in the copies or is it destroyed with the original? Is a human being just the atoms that constitute it plus the information determining its structure? If the human person were just software running on hardware constituted by the brain, then it may be possible to build a copy of the hardware and transfer the software, but the problem of the copies would still be there.

Besides, teleportation would require huge quantities of energy to transmit the enormous quantity of information needed, particularly if long distances are involved. Despite all these problems, research on quantum teleportation continues and a photon has been successfully teleported. The problem of the original also seems to have been solved: to extract the information needed to reconstruct the object, the original must be destroyed in the process. The teleportation of an elementary particle may be possible, but the teleportation of atoms and, above all, macroscopic objects is a completely different issue and is probably a dream that will remain so forever.

[12] L.M. Krauss, *The Physics of Star Trek*, Longanesi, Milano, 1996.

FIGURE 5.7 A symbolic image of the BPP program: a futuristic spaceship with warp propulsion. (NASA image.)

Even if wormholes and warp drives remind us of the tricks of science fiction, they are just very hypothetical ideas in a field of which we know very little. It is likely that if humans will one day succeed in traveling at speeds higher than that of light, they will do so by exploiting physical principles and phenomena that are as yet completely unknown. Such theoretical and technological knowledge has yet to be developed, and the only thing we can do is investigate all unknown phenomena with an open and critical mind.

We are therefore still very far from being able to give a positive answer to the possibility of interstellar journeys at speed higher than that of light.[13] Even if there is no certainty that, in the end, the efforts in this direction will lead to anything serious, whether in the immediate or very distant future, we can try to imagine the consequences of the possibility of traveling at interstellar distances in times that may be measured in hours and days rather than decades or centuries.

Even if radically new technologies permitted fast interstellar travel, they would nevertheless be long and expensive; the trick of allowing us to

---

[13] Actually, since we have the certainty that no object can move at speeds greater than that of light, we should say "with a travel time shorter than that of light."

travel at superluminal speed, for instance, will probably only be used at a great distance from the targeted celestial bodies, and the parts of the journey inside the origin and inside the destination system will be performed at a lower speed. Moreover, it might be necessary to limit the speed owing to the risk of encountering micrometeorites or other, even larger, bodies. Also, tens of days might be required, and perhaps much more, to leave one star system and enter another. Interstellar journeys may not take years but, as a minimum, will at least take many days.

This situation reminds us of the nineteenth century, when all inhabited zones of our planet could be reached, but the long journeys required were too expensive for most people to afford. However, these communication difficulties did not deter political empires from building communities and maintaining communication with their new territories on various continents.

In a scenario of this type, interstellar commerce cannot be entirely ruled out, but it is likely that there will not be many commodities whose value justifies transportation at such long distances. In the past, this was the case with spices, whose value was such that a shipload was worth several times the vessel transporting it. In addition to the possibility of commercial relationships, the political relationships between star systems are in the realm of possibilities, with the possible consequences of small and large conflicts.

If humankind colonizes a number of star systems, the situation of isolation described above will be much mitigated, although the instantaneous communications to which we are now accustomed, and which constitute the essence of the global village, will never be possible. If the human expansion outside the Solar System occurs with relatively slow spaceships or space arks, or with faster but nevertheless nonsuperluminal ships, then any method of traveling faster than light will only be discovered much later; and after the various communities had already consolidated a tradition of autonomous life, the unifying tendencies that new technologies may introduce could produce potentially conflicting situations. If, on the other hand, expansion can be performed directly with the use of superluminal spaceships, the process of colonization would be completely different, and no rapid initial differentiation would result from the isolation of the various communities.

The pace of human expansion in space and the characteristics of the planets orbiting the stars closest to us will have a strong influence on the type of civilization that results. If interstellar flight remains impossible for centuries, or even millennia, humans will have to adapt to the very hostile environments of the planets of the Solar System. In this case the pressures to adapt human beings to the environment will be very strong. The same will happen if the nearest stars have no habitable planets. If terrestrial-type planets can be found that are habitable without too many major difficulties, and can be reached for colonization within a reasonable time, it might be easier to contain the differentiation of humankind into different species.

The encounter with alien living beings and with extraterrestrial intelligence will occur in scenarios of this type. As seen in the preceding paragraph, if superluminal travel will forever remain impossible, the encounter will likely be of importance to a very small part of the human species, while for all the others the consequences will not be much different from those that could follow the success of SETI, since for them it will be just radio contact with a distant world.

The consequences of the encounter with other intelligences will be much more important if one of the two sides, or both, can travel or exchange messages at a speed greater than the speed of light. The exchange of messages could occur in both directions, and as it would be an actual interaction, it could facilitate the possibility of interpreting the messages from each other. From a scientific point of view, it would be possible to exchange information, even where the two intelligences were very different from each other. If the intelligent species were similar enough to live without too much discomfort in the same environment, it might even be possible to reach an actual face-to-face interaction without the need of interposed technological systems (audio and video connection, for instance, or other means we currently cannot even imagine), and therefore we could become even better acquainted.

True political relationships, with something similar to the exchange of ambassadors (intending such terms in the most general sense), might be reached and even economic relationships. There is no doubt that a scenario of this kind, however, could cause great risks: if a species had hegemonic or hostile intentions, an extremely dangerous situation could follow. Science fiction has accustomed us to scenarios of hostility or even of actual war between different species or members of the same species in interstellar scenarios. Rapid contact at very large distances could make such scenarios not only literary inventions but actual possibilities, and therefore any initiative for contact should be considered with great caution.

EPILOGUE

TODAY it is still impossible to give a final answer to the question "Are we alone in the Universe?"

The answers science has given in the past have been much different according to the times and the cultural trends but also the discipline professed by the scientist to whom the question was addressed. Biologists, for instance, have always been much more skeptical, while astronomers have the tendency to think that life is more or less widespread in the Universe.

There are three fundamental questions that can be defined as *questions on existence*.

The first concerns the existence of forms of life outside planet Earth. Nowadays there is a substantial agreement to give a positive answer, and a number of scientists think that the studies on some meteorites, particularly on ALH84001, coming from Mars, supplied the required evidence. Some scientists think that some traces of life were also found on other meteorites not coming from Mars. Life on Earth developed in such a short time after the formation of our planet as to suggest that there should be some mechanisms causing matter to organize in the form of living organisms. If they actually exist, the chances that life is widespread in the Universe are much greater.

The second question concerns the existence of complex life or, as it is often called, of animal life. This term is perhaps too highly influenced by what we know of terrestrial life: it is possible that complex extraterrestrial life cannot be assimilated to animal life, to vegetable life, or to anything we know. The answers to this question are much more differentiated, and while perhaps the prevailing opinion is that complex life is common in the

261

Universe, there is a view, synthesized in the so-called rare Earth hypothesis, that advocates the idea that Earth is, from this point of view, unique. Complex life, as we know it, should be detectable from the characteristics of the atmosphere of planets, since living beings produced, and maintain on Earth, an atmosphere rich in oxygen that could not exist without them. This consideration has also important implications for astronautics: to be habitable by Earthly life-forms, a planet must already be inhabited by complex forms of life of the same kind. To live on all other planets humans must either build artificial environments, as on the Moon or on Mars, or terraform the planet. Then, if the rare Earth hypothesis holds, the expansion of humankind into space will be more difficult.

The third question deals with intelligent and conscious life. We were never able to forward a general definition for intelligence or consciousness (it is very difficult even to give a general definition of what we mean by life), and we do not really know whether the two things are necessarily connected. Perhaps also for this reason the answers are much more differentiated. They span from the certainty that intelligence is the normal outcome of evolution of life to the opinion that intelligence constitutes a momentary anomaly that occurred on Earth and that soon will be corrected with the disappearance of the human species, perhaps caused by humans themselves.

Although it is likely that intelligent life is frequent enough to allow contacts among civilizations flourishing on planets orbiting different stars, we must also consider the possibility that intelligent beings are very rare, so rare that we might never come in contact with them or even that we might never reach the certainty that they exist. From all points of view, it is exactly as if we were alone in a Universe, perhaps rich with life, but not intelligent and conscious. In this case our responsibility would be even greater, as we could look at ourselves as the keepers of at least our part of the Universe.

Although the possible answers to these three fundamental questions are discussed in a large number of scientific conferences and lectures for the general public, and in scientific papers and popular science books, theoretical science will never be able to assess without some form of doubt the existence of extraterrestrial life or extraterrestrial intelligence. A demonstration of its incompatibility with some basic scientific principles could give us the certainty (for what scientific certainties are worth, since scientific statements must be subject to the possibility of a falsification) that it does not exist, while a proof of its possibility could not prove that it actually exists.

Experimental science, on the contrary—owing to the progress of its instruments—could prove that extraterrestrial life (by discovering, for instance, that forms of life, whose origin is surely independent from that of terrestrial organisms, exist or existed on Mars) or extraterrestrial intelligence (by receiving an artificial message from space) exists, while the lack of any experimental evidence will never supply the evidence that it does not exist. Very likely any experimental evidence will be much more difficult to obtain and be more controversial than is considered today, and it is not unlikely that we will continue for a long time to find clues and perhaps some evidence that is not necessarily conclusive. It is possible that it will not be sufficient to send automatic probes to Mars and that a crewed scientific expedition will have a hard and long job to find the required irrefutable evidence.

There is no doubt that to finally give an answer to any of the questions on the existence of extraterrestrial life, the criterion requiring exceptional evidence to support an exceptional discovery must be strictly applied.

If the questions on the existence of extraterrestrial life are answered in the affirmative, an almost endless number of other questions regarding the characteristics of this life and the possible relationships it may have with us, will be asked:

- Does a general scheme exist, mostly due to the uniformity of the laws of physics, causing all possible forms of life to display a certain uniformity?
- Is life necessarily based on carbon, on amino acids (always the same, with identical chirality?), and on DNA?
- Is cellular structure needed for life?
- Are complex living beings necessarily eukaryotes and is eukaryogenesis a necessary step in the development of life?

The questions multiply:

- Does it make sense to speak of convergent evolution in the case of life-forms that evolved in a completely independent way?
- How likely is it that, if intelligent beings evolved on other planets, the general configuration of their body is, at least in a general sense, humanoid?
- Much more importantly, can their mental structure be such to allow at least a minimum of mutual comprehension, or will they be so alien to us, in the most complete sense of the term, as to remain forever incomprehensible?
- Can we avoid turning absolute incomprehensibility into hostility?

The questions of this second type, those regarding more *how* than *if*, are endless and permit various answers with an infinity of distinctions. The thousand hypotheses that have been reported in this book are only attempts to give some answers, perhaps formulated more for the pleasure of imagining what reality may be than in the hope of getting close to a truth that, as Galileo anticipated five centuries ago, will probably be beyond even the wildest imagination.

The effort of imagining other living beings and other intelligent creatures, and different ways of being intelligent, is not a sterile exercise; the result may change our views of the Universe and of life—no more an indifferent or even hostile Universe but, on the contrary, a Universe that promotes the organization of matter, creating those admirable structures on which life, and then intelligence, are based. And life can be no more interpreted as a fragile and momentary anomaly, but as the main route of cosmic evolution.

But this intellectual exercise in imagining other worlds would just become a game of fantasy if it does not stimulate actions aimed at verifying, through experimental science, the hypotheses that are proposed.

For the first time in history humankind has the means to try to give a final answer to these questions. Technology is preparing instruments that, in the future, may be able to give this answer, both through exploration, first by robots and then directly by humans, of celestial bodies at increasingly large distances, and through astronomic observations in the whole spectrum, from gamma-rays and x-rays to radio waves, that may bring us signals that have been broadcast by extraterrestrial intelligences.

As always, the theoretical elaboration and the refinement of the experimental techniques are complementary. Both aspects are full of difficulties, but the importance of the result is too great to abandon. This search will also lead, as a side benefit, to a better understanding of life on Earth and of human nature; even if a final result will not be reached within a predictable time, very important results will be obtained in these areas.

But the study of the role of life and intelligence in the Universe goes well beyond the boundaries of science and enters the fields of philosophy and religion.

Any philosophy that intends to give an interpretation of the world and an understanding of the role that humankind must play in it cannot avoid considering this aspect. Likewise, the existence of life, and above all of intelligent and conscious life, distant from our planet has strong religious implications.

From these points of view we cannot avoid extending the concepts that we have developed on our planet to a wider environment. Life and, above all, intelligent life that in the future may be discovered distant from our planet, will need to be considered in the same way as terrestrial life, independently from the forms it takes and from its biochemical bases. There is no doubt that it could be difficult to apply this general consideration to actual cases, in which it may be even doubtful whether certain structures can be considered as living or whether a certain form of life may be considered intelligent. Even greater difficulties may be found in understanding whether a form of life is conscious: intelligence and consciousness could be present in a different degree instead of being, as we have the tendency to think, qualities that can only be either present or absent. If consciousness is really linked with complexity, not only as a metaphor to understand the origin of intelligent life but owing to an actual law of nature, we could meet beings of intermediate complexity that cannot be defined clearly.

In spite of these difficulties, humankind must be ready to recognize as peers all the other intelligent and conscious beings that it will discover or will meet in its expansion in the Universe. In the absence of a more general term synthesizing the essence of a conscious and intelligent living being, the meaning of human (man and woman, provided that the distinction in genders is applicable) must be extended to include all intelligent species. The discovery of extraterrestrial intelligence will not, therefore, be an encounter between humans and aliens, but between humans from planet Earth and humans living in the depths of space.

APPENDIX A

EXTRASOLAR PLANETS (DECEMBER 23, 2005)[1]

THE total number of extrasolar planets orbiting normal stars (i.e., main sequence stars) discovered by the end of 2005 is 170. They belong to 146 planetary systems, eighteen of which have more than one planet.

They are reported in the table below; the name of the planet is given by the name of the star, followed by the letter *b* for the first planet discovered, *c* for the second, *d* for the third, and so on.

The mass is actually the product $m \times \sin(i)$ (see text) measured in multiples of the mass of Jupiter. The period is expressed in days, the semi-major axis of the orbit in astronomical units (AU), and the distance of the star from the Sun in parsecs. The table is ordered by increasing distance of the planetary system from the Sun.

Name of planet	Mass	Period	Orbit semi-major axis	Eccentricity	Star distance
Epsilon Eridani *b*	0.86	2502.1	3.3	0.61	3.2
Gliese 876 *b*	1.94	60.94	0.21	0.02	4.7
Gliese 876 *c*	0.56	30.1	0.13	0.27	4.7
Gliese 876 *d*	0.02	1.94	0.02	0	4.7
Gl 581 *b*	0.06	5.37	0.04	0	6.3

[1] http://vo.obspm.fr/exoplanetes/encyclo/encycl.html.

Name of planet	Mass	Period	Orbit semi-major axis	Eccentricity	Star distance
GJ 436 b	0.07	2.64	0.03	0.12	10.2
Gl 86 b	4.01	15.77	0.11	0.05	11
HD 3651 b	0.2	62.23	0.28	0.63	11
Gamma Cephei b	1.59	902.26	2.03	0.2	11.8
HD 147513 b	1	540.4	1.26	0.52	12.9
47 Uma b	2.54	1089	2.09	0.06	13.3
47 Uma c	0.76	2594	3.73	0.1	13.3
55 Cnc b	0.78	14.67	0.12	0.02	13.4
55 Cnc c	0.22	43.93	0.24	0.44	13.4
55 Cnc d	3.92	4517.4	5.26	0.33	13.4
55 Cnc e	0.05	2.81	0.04	0.17	13.4
Ups And b	0.69	4.62	0.06	0.01	13.5
Ups And c	1.89	241.5	0.83	0.28	13.5
Ups And d	3.75	1284	2.53	0.27	13.5
51 Peg b	0.47	4.23	0.05	0	14.7
Tau Boo b	3.9	3.31	0.05		15
HD 160691 b	1.67	654.5	1.5	0.31	15.3
HD 160691 c	3.1	2986	4.17	0.57	15.3
HD 160691 d	0.04	9.55	0.09	0	15.3
HR 810 b	1.94	311.29	0.91	0.24	15.5
HD 190360 c	0.06	17.1	0.13	0.01	15.9
HD 190360 b	1.5	2891	3.92	0.36	15.9
HD 128311 b	2.18	448.6	1.1	0.25	16.6
HD 128311 c	3.21	919	1.76	0.17	16.6
rho CrB b	1.04	39.85	0.22	0.04	16.7
HD 10647 b	0.91	1040	2.1	0.18	17.3
GJ 3021 b	3.32	133.82	0.49	0.51	17.6
HD 99492 b	0.12	17.04	0.12	0.05	18
14 Her b	4.74	1796.4	2.8	0.34	18.1
HD 27442 b	1.28	423.84	1.18	0.07	18.1
HD 189733 b	1.15	2.22	0.03	0	19.3
HD 192263 b	0.72	24.35	0.15	0	19.9
HD 195019 b	3.43	18.3	0.14	0.05	20
HD 39091 b	10.35	2063.82	3.29	0.62	20.6
HD 142 b	1	337.11	0.98	0.38	20.6

Appendix A

Name of planet	Mass	Period	Orbit semi-major axis	Eccentricity	Star distance
HD 33564 b	9.1	388	1.1	0.34	21
16 Cyg B b	1.69	798.94	1.67	0.67	21.4
HD 4308 b	0.05	15.56	0.11	0	21.9
70 Vir b	7.44	116.69	0.48	0.4	22
HD 114783 b	0.99	501	1.2	0.1	22
HD 210277 b	1.24	435.6	1.1	0.45	22
HD 19994 b	2	454	1.3	0.2	22.4
HD 134987 b	1.58	260	0.78	0.24	25
HD 216437 b	2.1	1294	2.7	0.34	26.5
HD 179949 b	0.98	3.09	0.04	0.05	27
HD 20367 b	1.07	500	1.25	0.23	27
HD 150706 b	1	264	0.82	0.38	27.2
HD 82943 b	1.75	441.2	1.19	0.22	27.5
HD 82943 c	2.01	219	0.75	0.36	27.5
HD 52265 b	1.13	118.96	0.49	0.29	28
HD 114762 b	11.02	83.89	0.3	0.34	28
HD 114386 b	0.99	872	1.62	0.28	28
HD 33636 b	9.28	2447.29	3.56	0.53	28.7
HD 93083 b	0.37	143.58	0.48	0.14	28.9
HD 75289 b	0.42	3.51	0.05	0.05	28.9
HD 70642 b	2	2231	3.3	0.1	29
HD 111232 b	6.8	1143	1.97	0.2	29
HD 130322 b	1.08	10.72	0.09	0.05	30
HD 10697 b	6.12	1077.91	2.13	0.11	30
HD 101930 b	0.3	70.46	0.3	0.11	30.5
HD 50554 b	4.9	1279	2.38	0.42	31
HD 162020 b	13.75	8.43	0.07	0.28	31.3
HIP 75458 b	8.64	550.65	1.34	0.71	31.5
HD 81040 b	6.86	1001.7	1.94	0.53	32.6
HD 92788 b	3.86	377.7	0.97	0.27	32.8
HD 168443 b	7.2	58.12	0.29	0.53	33
HD 168443 c	17.1	1739.5	2.87	0.23	33
HD 37124 b	0.61	154.46	0.53	0.06	33
HD 37124 c	0.6	843.6	1.64	0.14	33
HD 37124 d	0.66	2295	3.19	0.2	33
HD 89307 b	2.73	3090	4.15	0.27	33
HD 117207 b	2.06	2627.08	3.78	0.16	33
HD 196885 b	1.84	386	1.12	0.3	33

Appendix A

Name of planet	Mass	Period	Orbit semi-major axis	Eccentricity	Star distance
HD 40979 b	3.32	267.2	0.81	0.23	33.3
HD 216435 b	1.49	1442.92	2.7	0.34	33.3
HD 46375 b	0.25	3.02	0.04	0.04	33.4
HD 141937 b	9.7	653.22	1.52	0.41	33.5
HD 4208 b	0.8	812.2	1.67	0.05	33.9
HD 11964 b	0.11	37.82	0.23	0.15	34
HD 11964 c	0.7	1940	3.17	0.3	34
HD 142415 b	1.62	386.3	1.05	0.5	34.2
HD 65216 b	1.21	613.1	1.37	0.41	34.3
HD 23079 b	2.61	738.46	1.65	0.1	34.8
HD 114729 b	0.82	1131.48	2.08	0.31	35
HD 63454 b	0.38	2.82	0.04	0	35.8
HD 142022 A b	4.4	1923	2.8	0.57	35.9
HD 16141 b	0.23	75.56	0.35	0.21	35.9
HD 169830 b	2.88	225.62	0.81	0.31	36.3
HD 169830 c	4.04	2102	3.6	0.33	36.3
HD 73256 b	1.87	2.55	0.04	0.03	36.5
HD 217107 b	1.37	7.13	0.07	0.13	37
HD 217107 c	2.1	3150	4.3	0.55	37
HD 12661 b	2.3	263.6	0.83	0.35	37.2
HD 12661 c	1.57	1444.5	2.56	0.2	37.2
HD 106252 b	6.81	1500	2.61	0.54	37.4
HD 216770 b	0.65	118.45	0.46	0.37	38
HD 117618 b	0.19	52.2	0.28	0.39	38
HD 108147 b	0.4	10.9	0.1	0.5	38.6
HD 28185 b	5.7	383	1.03	0.07	39.4
HD 89744 b	7.99	256.61	0.89	0.67	40
HD 6434 b	0.48	22.09	0.15	0.3	40.3
HD 49674 b	0.11	4.95	0.06	0.16	40.7
HD 213240 b	4.5	951	2.03	0.45	40.8
HD 222582 b	5.11	572	1.35	0.76	42
HD 102117 b	0.14	20.67	0.15	0.06	42
HD 27894 b	0.62	17.99	0.12	0.05	42.4
HD 38529 b	0.78	14.31	0.13	0.29	42.4
HD 38529 c	12.7	2174.3	3.68	0.36	42.4
HD 41004 A b	2.3	655	1.31	0.39	42.5

Appendix A

Name of planet	Mass	Period	Orbit semi-major axis	Eccentricity	Star distance
HD 37605 b	2.3	55	0.25	0.68	42.9
HD 168746 b	0.23	6.4	0.07	0.08	43.1
HD 83443 b	0.41	2.99	0.04	0.08	43.5
HD 8574 b	2.23	228.8	0.76	0.4	44.2
HD 121504 b	0.89	64.6	0.32	0.13	44.4
HD 188753A b	1.14	3.35	0.04	0	44.8
HD 219449 b	2.9	182	0.3		45
HD 208487 b	0.45	123	0.49	0.32	45
HD 202206 b	17.4	255.87	0.83	0.44	46.3
HD 202206 c	2.44	1383.4	2.55	0.27	46.3
HD 178911 B b	6.29	71.49	0.32	0.12	46.7
HD 196050 b	3	1289	2.5	0.28	46.9
HD 209458 b	0.69	3.52	0.05	0.07	47
HD 50499 b	1.71	2582.7	3.86	0.23	47.3
HD 34445 b	0.58	126	0.51	0.4	48
HD 45350 b	0.98	890.76	1.77	0.78	49
HD 187123 b	0.52	3.1	0.04	0.03	50
HD 330075 b	0.76	3.37	0.04	0	50.2
HD 72659 b	2.96	3177.4	4.16	0.2	51.4
HD 23596 b	7.19	1558	2.72	0.31	52
HD 136118 b	11.9	1209	2.3	0.37	52.3
HD 188015 b	1.26	456.46	1.19	0.15	52.6
HD 212301 b	0.05	2.46	0.04	0	52.7
HD 183263 b	3.69	634.23	1.52	0.38	53
HD 2638 b	0.48	3.44	0.04	0	53.7
HD 30177 b	9.17	2819.65	3.86	0.3	55
HD 68988 b	1.9	6.28	0.07	0.14	58
HD 80606 b	3.41	111.78	0.44	0.93	58.4
HD 109749 b	0.28	5.24	0.06	0.01	59
HD 177830 b	1.28	391	1	0.43	59
HD 76700 b	0.2	3.97	0.05	0.13	59.7
HD 149143 b	1.33	4.07	0.05	0.02	63
HD 74156 b	1.86	51.64	0.29	0.64	64.6
HD 74156 c	6.17	2025	3.4	0.58	64.6
HD 190228 b	4.99	1127	2.31	0.43	66.1
HD 11977 b	6.54	711	1.93	0.4	66.5
HD 108874 c	1.02	1605.8	2.68	0.25	68.5

Name of planet	Mass	Period	Orbit semi-major axis	Eccentricity	Star distance
HD 108874 *b*	1.36	395.4	1.05	0.07	68.5
HD 154857 *b*	1.8	398.5	1.11	0.51	68.5
HD 88133 *b*	0.22	3.41	0.05	0.11	74.5
HD 4203 *b*	1.65	400.94	1.09	0.46	77.5
HD 149026 *b*	0.36	2.88	0.04	0	78.9
HD 118203 *b*	2.13	6.13	0.07	0.31	88.6
HD 2039 *b*	4.85	1192.58	2.19	0.68	89.8
HD 59686 *b*	5.25	303	0.91	0	92
HD 73526 *b*	3	190.5	0.66	0.34	99
HD 104985 *b*	6.3	198.2	0.78	0.03	102
HD 47536 *b*	4.96	712.13	1.61	0.2	123
TrES-1 *b*	0.61	3.03	0.04	0.14	157
HD 13189 *b*	14	471.6	1.85	0.28	185
OGLE-TR-56 *b*	1.45	1.21	0.02	0	1,500
OGLE-TR-113 *b*	1.35	1.43	0.02	0	1,500
OGLE-TR-132 *b*	1.19	1.69	0.03	0	1,500
OGLE-TR-10 *b*	0.54	3.1	0.04	0	1,500
OGLE-TR-111 *b*	0.53	4.02	0.05	0	1,500
BD-10 3166 *b*	0.48	3.49	0.05	0.07	—

Four planets were discovered around two pulsars:

Name of planet	Mass	Period	Orbit semi-major axis	Eccentricity	Star distance
PSR 1257+12 *b*	0.0000629	25.26	0.19	0.0	300
PSR 1257+12 *c*	0.0135	66.54	0.36	0.0186	300
PSR 1257+12 *d*	0.0122	98.21	0.46	0.0252	300
PSR B1620-26 *b*	2.5	100 (y)	23	—	3,800

Apart from the planets listed above, a certain number of planets were discovered, but the discovery was then not confirmed or rejected. They are not listed here.

APPENDIX B

Following the very detailed survey by A.M. MacRoberts published on *Sky & Telescope*,[2] a condensed survey of the current radioastronomic SETI projects is presented below.

It has been said that SETI is like searching for a needle in a cosmic haystack; an idea of the size of the haystack and of the fraction searched by the present programs is given by Figure B.1. The search space is represented by the tridimensional plot, whose three axes are the frequency (in GHz), the fraction of the sky, and the sensitivity. The latter axis is scaled using an arbitrary unit: the number of stars that can be examined in a given direction for an alien transmitter of a given power. It is clear that each of the projects presently under way has a weakness in at least one of the axes, and the total volume of the haystack searched is a small fraction of the total.

Project Phoenix

Project Phoenix was run by the SETI Institute of Mountain View, California. It ended in March 2004.

---

[2] http://www.skyandtelescope.com/printable/resources/seti/article_251.asp

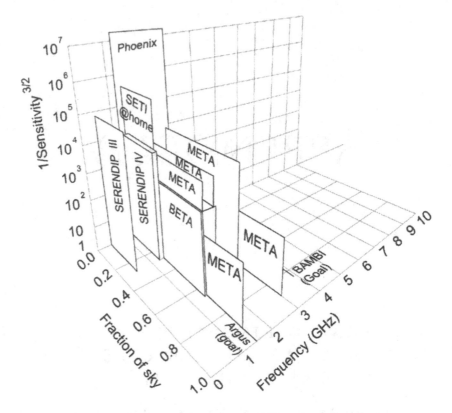

FIGURE B.1 *Search space of the radioastronomic SETI projects presently under way.*

The project was started with private funding to recover the work already performed by NASA when, in 1993, the very ambitious NASA SETI research was canceled by the U.S. Congress. It owes its name to rising from its own ashes. Funding for the SETI Institute comes from hi-tech companies and many individuals.

The targets were a number of stars chosen for their type (mostly solar type), distance (less than 150 light-years), and age (more than three billion years). A total of 800 stars were examined on more than two billion channels between 1.2 and 3 GHz with a resolution of 0.7 Hz. The equipment for Project Phoenix is installed in a truck trailer that can be moved close to different radio telescopes to perform its task. In the last six years of the program it operated from Arecibo radio telescope, were it was granted 5 percent of the observation time.

As the project has ended, the equipment is being upgraded, and the project will start again using the new Allen Telescope Array (ATA) that the SETI Institute is building in northern California.

Appendix B

Project SERENDIP

Project SERENDIP (Search for Extraterrestrial Radio Emissions from Nearby Developed Intelligent Populations) is run by researchers of the University of California at Berkeley. Its main aim is to overcome the biggest problem of all SETI programs: obtaining precious observation time at the various radio telescopes. More than 25 years ago, the researchers of University of California at Berkeley thought of putting a SETI receiver in "piggyback" onto a radio telescope and working while the main receiver was doing its main tasks without interference. The program started in 1978 with an early device and continued with an updated system. At present the SERENDIP IV receiver is installed on the Arecibo radio telescope. It can receive 168 million channels simultaneously, each 0.6 Hz wide, in a 100-MHz band centered on the frequency of hydrogen (see Figure 4.6).

Operating from Arecibo, SERENDIP IV is limited to a zone of the sky between declination 2° and 35°, that is, about 30 percent of the whole sky. Each point in that zone is repeatedly covered in time.

The next generation receiver, SERENDIP V, is almost ready and should enter service soon, again from Arecibo. It will monitor a number of channels three times larger than the previous instrument.

SETI@home

SETI@home started in 1999, thanks to the Planetary Society, which contributed $50,000 from the Carl Sagan Memorial Fund for the Future to a project by David Gedye, a computer scientist from Seattle, who, since 1994, had been trying to organize a network of volunteers to analyze, with their home computers, the signals from radio telescopes.

The project is linked with SERENDIP, and aims to solve one of the larger problems of the latter. Even the very powerful supercomputer purposely designed for this project must limit itself to looking for the simplest signals with predetermined characteristics. The data regarding the central 2.5-MHz band (out of the 100-MHz band of SERENDIP IV)— that is, those closest to the hydrogen line—are subdivided into work units of 354 kilobytes and sent to the connected computers to be elaborated. Depending on the speed of the computer, the time to complete the job with a work unit is between 3 and 40 hours.

The data so dealt with can be examined in much greater detail than is possible with the standard SERENDIP computer, increasing the sensitivity of the search both for signal intensity (it can detect signals 10 times weaker) and signal bandwidth (down to an 0.075-Hz bandwidth).

Apart from exploiting the very large computing power that would otherwise be wasted, SETI@home is very useful as a way of making many people aware of SETI research.

Southern SERENDIP

Since 1998 a system very similar to SERENDIP III started operating in piggyback on the 64-m radio telescope at Parkes Observatory in Australia. The SERENDIP III receiver was upgraded to receive 58.8 million channels, each 0.6 Hz wide for a total bandwidth of 36 MHz centered on a frequency that can be adjusted between 1.2 and 1.5 GHz. The search is conducted by the SETI Australia Center, of the University of Western Sidney Macarthur. Its sensitivity is much smaller than that of the SERENDIP program based at Arecibo, owing to the fact that the dish has a diameter five times smaller (and hence its collecting area is twenty-five times smaller), but it can be oriented to survey a large part of the sky, mostly in the Southern Hemisphere and out of the coverage of Arecibo.

Project BETA

Project BETA (Billion-channel Extra-Terrestrial Assay) was run from 1995 to 1999 by Paul Horowitz and his graduate students of Harvard University using a 26-m dish in the town of Cambridge, Massachusetts. It was sponsored by the Planetary Society and private donors. The project was stopped in 1999 when one of the gears of the antenna support (controlling the right ascension) failed during a wind storm and the dish crashed to the ground. The Planetary Society raised funds to repair the radio telescope, but the damage was found to be worse than expected and the repairs were put on hold. In 2005 Horowitz said in an interview with *Sky & Telescope*: "BETA is dead."

BETA scanned a wide frequency band from 1.40 to 1.72 GHz, with a resolution of 0.5 Hz, surveying as much as 68 percent of the sky.

Project META II

META (Million-channel Extra-Terrestrial Assay) project was the fore-runner of the BETA project. The META equipment was duplicated and installed on two antennas located near Buenos Aires and run by the Argentine Institute of Radioastronomy with funding from the Planetary Society. The project, directed by Guillermo Lemarchand, monitors 8.4 million very narrow channels, each 0.05 Hz wide, close to the hydrogen line and its second harmonics. It can monitor the southern sky, between declination −90° and −10°.

SETI Italia

A 24-million-channel SERENDIP IV analyzer was piggybacked to the 32-m dish of the Istituto di Radioastronomia of the Italian National Research Council at Medicina close to Bologna, Italy. The aim of the search, directed by Stelio Montebugnoli, is to scan at least 50 percent of the sky observable from Medicina, covering a 15-MHz bandwidth at 0.6-Hz resolution.

A new more powerful analyzer, with 64 million channels and the use of the new radio telescope of Noto (Sicily), are planned for the near future.

Project Argus and Amateur SETI

The SETI League, directed by Paul Shuch, has the ambitious plan to coordinate the many amateur SETI projects worldwide through its Argus Project. The goal is that of coordinating 5000 small stations worldwide, in such a way as to keep the whole sky under continuous observation. In June 2004, however, Project Argus had only coordinated about 130 participants.

As part of the Amateur SETI, the Project BAMBI (Bob and Mike's Big Investment) operated two small dishes 1000 miles apart in California and Colorado to screen out local interference. Another is the SETI Research & Community Development Institute in Australia that coordinates many amateurs building small antennas. It is working to obtain the rights to use the huge dish (33.5-m diameter) at Carnavon.

Project MSTAR

The MSTAR (Montecorvino SETI Array) is a project between professional and amateur SETI. The amateur astronomical observatory of Montecorvino (close to Salerno, Italy) is planning to build an array of dishes for SETI research under the direction of SETI professional researchers like Paul Shuch and Stelio Montebugnoli, using mainly public funding. The main interest of the project is that it is a radioastronomic observatory purposely built for SETI. Initially it will be based on an array of five dishes of 4.5 m and, if funding will allow, will be expanded to a twenty-five-dish array.

APPENDIX C

DECLARATION OF PRINCIPLES CONCERNING ACTIVITIES FOLLOWING DETECTION OF EXTRATERRESTRIAL INTELLIGENCE

We, the institutions and individuals participating in the search for extraterrestrial intelligence:

- recognizing that the search for extraterrestrial intelligence is an integral part of space exploration and is being undertaken for peaceful purposes and for the common interest of all mankind;
- inspired by the profound significance for mankind of detecting evidence of extraterrestrial intelligence, even though the probability of detection may be low;
- recalling the Treaty on Principles Governing the Activities of States in the Exploration and Use of Outer Space, Including the Moon and Other Celestial Bodies, which commits States Parties to that Treaty "to inform the Secretary General of the United Nations as well as the public and the international scientific community, to the greatest extent feasible and practicable, of the nature, conduct, locations, and results" of their space exploration activities (Article XI);
- recognizing that any initial detection may be incomplete or ambiguous and thus require careful examination as well as confirmation, and that it is essential to maintain the highest standards of scientific responsibility and credibility;

agree to observe the following principles for disseminating information about the detection of extraterrestrial intelligence:

1. Any individual, public or private research institution, or governmental agency that believes it has detected a signal from or other evidence of extraterrestrial intelligence (the discoverer) should seek to verify that the most plausible explanation for the evidence is the existence of extraterrestrial intelligence rather than some other natural phenomenon or anthropogenic phenomenon before making any public announcement. If the evidence cannot be confirmed as indicating the existence of extraterrestrial intelligence, the discoverer may disseminate the information as appropriate to the discovery of any unknown phenomenon.

2. Prior to making a public announcement that evidence of extraterrestrial intelligence has been detected, the discoverer should promptly inform all other observers or research organizations that are parties to this declaration, so that those other parties may seek to confirm the discovery by independent observations at other sites and so that a network can be established to enable continuous monitoring of the signal or phenomenon. Parties to this declaration should not make any public announcement of this information until it is determined whether this information is or is not credible evidence of the existence of extraterrestrial intelligence. The discoverer should inform his/her or its relevant national authorities.

3. After concluding that the discovery appears to be credible evidence of extraterrestrial intelligence, and after informing other parties to this declaration, the discoverer should inform observers throughout the world through the Central Bureau for Astronomical Telegrams of the International Astronomical Union, and should inform the Secretary General of the United Nations in accordance with Article XI of the Treaty on Principles Governing the Activities of States in the Exploration and Use of Outer Space, Including the Moon and Other Bodies. Because of their demonstrated interest in and expertise concerning the question of the existence of extraterrestrial intelligence, the discoverer should simultaneously inform the following international institutions of the discovery and should provide them with all pertinent data and recorded information concerning the evidence: the International Telecommunication Union, the Committee on Space Research, of the International Council of Scientific Unions, the International Astronautical Federation, the International Academy of Astronautics, the International Institute of Space Law, Commission 51 of the International Astronomical Union, and Commission J of the International Radio Science Union.

4. A confirmed detection of extraterrestrial intelligence should be disseminated promptly, openly, and widely through scientific channels and public media, observing the procedures in this declaration. The discoverer should have the privilege of making the first public announcement.

5. All data necessary for confirmation of detection should be made available to the international scientific community through publications, meetings, conferences, and other appropriate means.

6. The discovery should be confirmed and monitored and any data bearing on the evidence of extraterrestrial intelligence should be recorded and stored permanently to the greatest extent feasible and practicable, in a form that will make it available for further analysis and interpretation. These recordings should be made available to the international institutions listed above and to members of the scientific community for further objective analysis and interpretation.

7. If the evidence of detection is in the form of electromagnetic signals, the parties to this declaration should seek international agreement to protect the appropriate frequencies by exercising procedures available through the International Telecommunication Union. Immediate notice should be sent to the Secretary General of the ITU in Geneva, who may include a request to minimize transmissions on the relevant frequencies in the Weekly Circular. The Secretariat, in conjunction with advice of the Union's Administrative Council, should explore the feasibility and utility of convening an Extraordinary Administrative Radio Conference to deal with the matter, subject to the opinions of the member Administrations of the ITU.

8. No response to a signal or other evidence of extraterrestrial intelligence should be sent until appropriate international consultations have taken place. The procedures for such consultations will be the subject of a separate agreement, declaration or arrangement.

9. The SETI Committee of the International Academy of Astronautics, in coordination with Commission 51 of the International Astronomical Union, will conduct a continuing review of procedures for the detection of extraterrestrial intelligence and the subsequent handling of the data. Should credible evidence of extraterrestrial intelligence be discovered, an international committee of scientists and other experts should be established to serve as a focal point for continuing analysis of all observational evidence collected in the aftermath of the discovery, and also to provide advice on the release of information to the public. This committee should be constituted from representatives of each of the international institutions listed above and such other members as

the committee may deem necessary. To facilitate the convocation of such a committee at some unknown time in the future, the SETI Committee of the International Academy of Astronautics should initiate and maintain a current list of willing representatives from each of the international institutions listed above, as well as other individuals with relevant skills, and should make that list continuously available through the Secretariat of the International Academy of Astronautics. The International Academy of Astronautics will act as the Depository for this declaration and will annually provide a current list of parties to all the parties to this declaration.

INDEX

Index